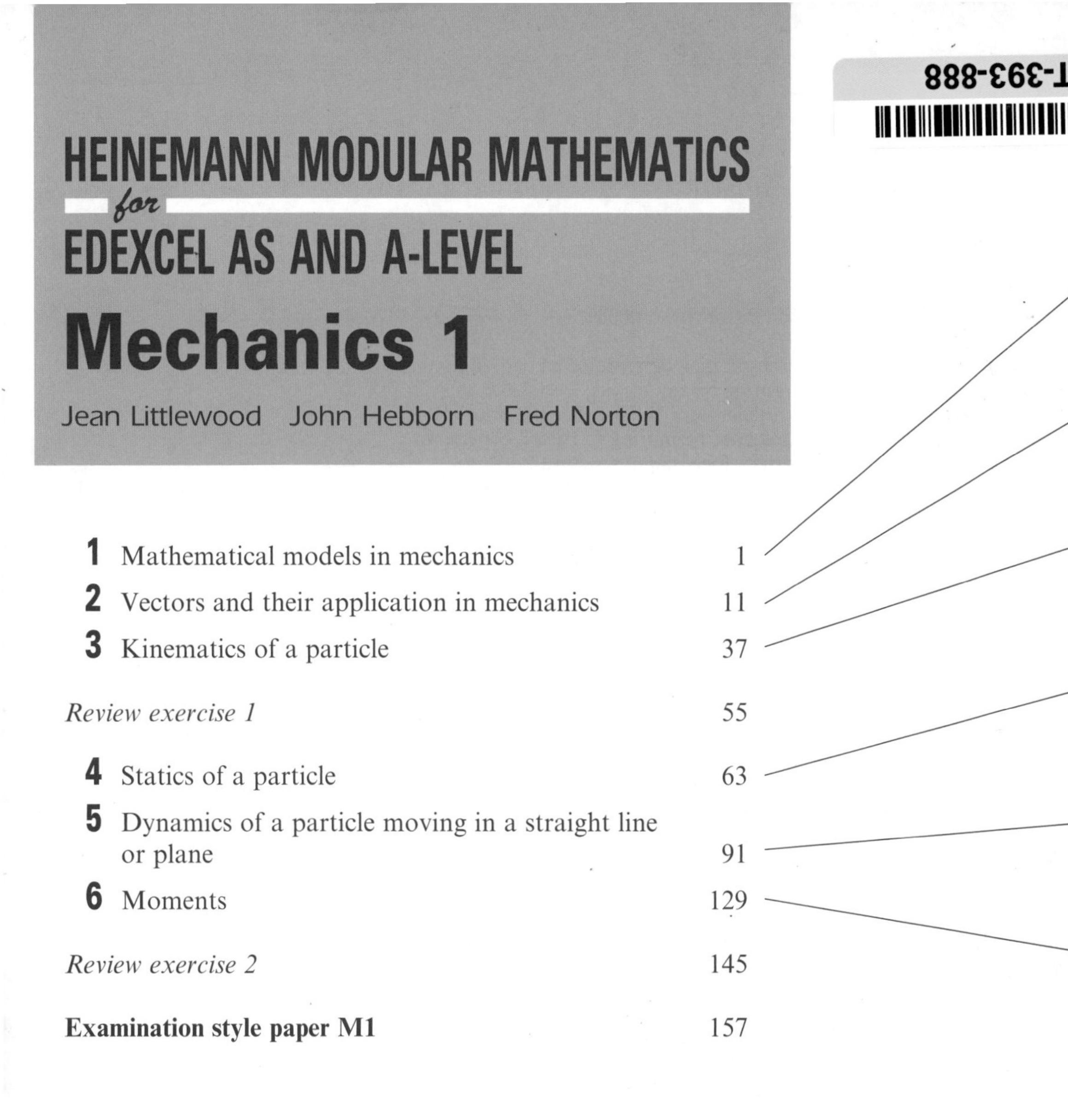

HEINEMANN MODULAR MATHEMATICS *for* EDEXCEL AS AND A-LEVEL

Mechanics 1

Jean Littlewood   John Hebborn   Fred Norton

Endorsed by edexcel

heinemann.co.uk

✓ Free online support

✓ Useful weblinks

✓ 24 hour online ordering

01865 888058

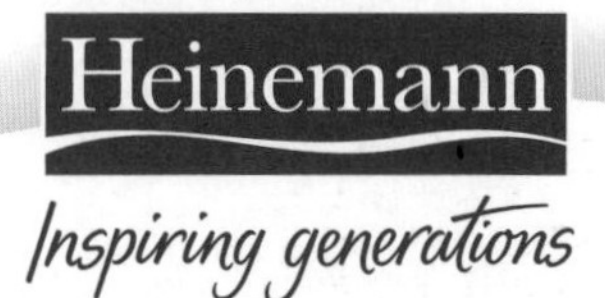

Heinemann Educational Publishers
Halley Court, Jordan Hill, Oxford OX2 8EJ
Part of Harcourt Education

Heinemann is the registered trademark of
Harcourt Education Limited

First published 2000

07 15

13-digit ISBN 978 0 435510 74 9

Cover design by Gecko Limited

Original design by Geoffrey Wadsley; additional design work by Jim Turner

Typeset and illustrated by Tech-Set Limited, Gateshead, Tyne and Wear

Printed in China by China Translation & Printing Services Ltd.

Acknowledgements:

The publisher's and authors' thanks are due to Edexcel for permission to reproduce questions from past examination papers. These are marked with an [E].

The answers have been provided by the authors and are not the responsibility of the examining board.

About this book

This book is designed to provide you with the best preparation possible for your Edexcel M1 exam. The series authors are senior examiners and exam moderators themselves and have a good understanding of Edexcel's requirements.

Use this **new edition** to prepare for the new 6-unit specification. Use the first edition (*Heinemann Modular Mathematics for London AS and A-level*) if you are preparing for the 4-module syllabus.

Finding your way around

To help you find your way around when you are studying and revising use the:

- **edge marks** (shown on the front page) – these help you to get to the right chapter quickly;
- **contents list** – this lists the headings that identify key syllabus ideas covered in the book so you can turn straight to them;
- **index** – if you need to find a topic the **bold** number shows where to find the main entry on a topic.

Remembering key ideas

We have provided clear explanations of the key ideas and techniques you need throughout the book. Key ideas you need to remember are listed in a **summary of key points** at the end of each chapter and marked like this in the chapters:

■ $\mathbf{v} = \mathbf{u} + \mathbf{a}t$

Exercises and exam questions

In this book questions are carefully graded so they increase in difficulty and gradually bring you up to exam standard.

- **past exam questions** are marked with an [E];
- **review exercises** on pages 55 and 145 help you practise answering questions from several areas of mathematics at once, as in the real exam;
- **exam style practice paper** on page 157 – this is designed to help you prepare for the exam itself;
- **answers** are included at the end of the book – use them to check your work.

Contents

Mathematical models in mechanics

1

1.1 What is mathematical modelling?

Physicists and engineers have used mathematics to solve problems in the real world for more than 200 years. Nowadays it is also used to solve problems in many other fields such as biology, economics, geography, medicine and psychology. Examples of problems which may be solved using mathematics include:

- estimating the height of the leaning tower of Pisa – without climbing it
- estimating the width of a river – without crossing it
- predicting the population of China in the year 2020 – without waiting until then
- predicting the effect of a 30 per cent reduction in income tax – without actually reducing the rate
- estimating the volume of blood inside someone's body – without bleeding him to death!

These problems and many others may be solved using a process called **mathematical modelling**. Mathematical modelling involves:

- **translating** a real world problem into a mathematical problem or **model**;
- **solving** the mathematical problem;
- **interpreting** the solution in terms of the real world.

The process can be shown as a diagram:

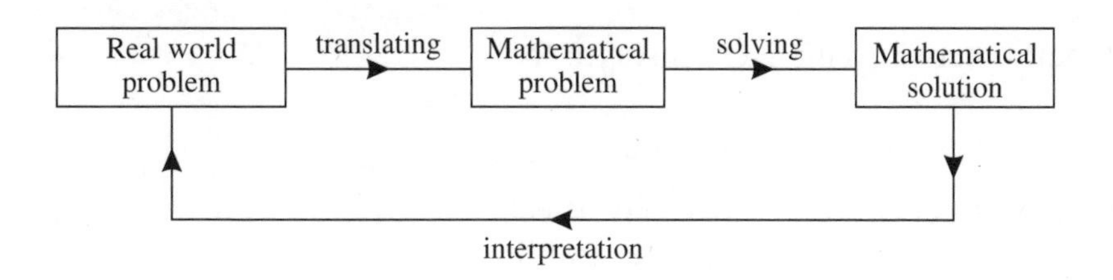

Real world problems can't often be translated perfectly into mathematical problems. Even when they can, the resulting mathematical problem may be so complex that it is not possible to solve it. So it is often necessary to simplify a real world problem into one which produces a solvable mathematical problem by concentrating on its essential features and ignoring the rest.

Sometimes the amount of simplification required may seem drastic. For example, to model the motion of the planets, the planets, and the Sun can be considered to be point masses and their size and structure ignored. Simplifications like this may be justified if the resulting mathematical model produces predictions for the motions of the planets that match up to the motions that are actually observed. If the model does not produce reliable predictions it may need to be changed in some way, usually by considering the assumptions made in the simplification. The modelling process can be shown in a diagram like this:

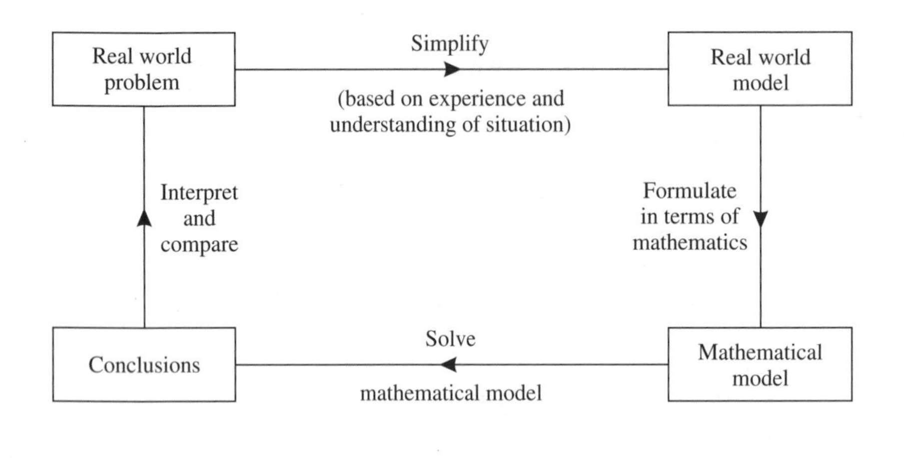

Example 1

A student is asked to estimate her travelling expenses when invited to an interview. She models the situation in a very simple way. First she looks up the distance d km in her road atlas. She finds from her car manual that her car should do k km per litre. She knows from petrol station prices that the cost of unleaded petrol is £c per litre.

Based on this data she estimates her travelling expenses to be £$\left(\frac{d}{k}\right)c$.

This is a simple model which ignores many factors that may affect the number of km per litre her car will actually travel: the traffic conditions, possible roadworks, the types of roads she travels on, the time of day she travels – rush hour or off peak. It also ignores other running costs of her car.

Using her experience she may then improve or refine her model, for example by changing the value of k or building in some of the factors neglected in her simple model.

Example 2

Here is a much more complex problem – how can we estimate the rate at which a country's population will grow? Changes in the way resources are used need to be planned well ahead, so governments need good estimates of how the population is likely to change. The real world problem is to explain how populations change, and to create a model to predict future changes.

The first model was put forward by Thomas Malthus as long ago as 1798. He suggested that the rate of change of a population of size N people is kN, where k is a positive constant. The mathematical solution of this problem produces the graph shown here.

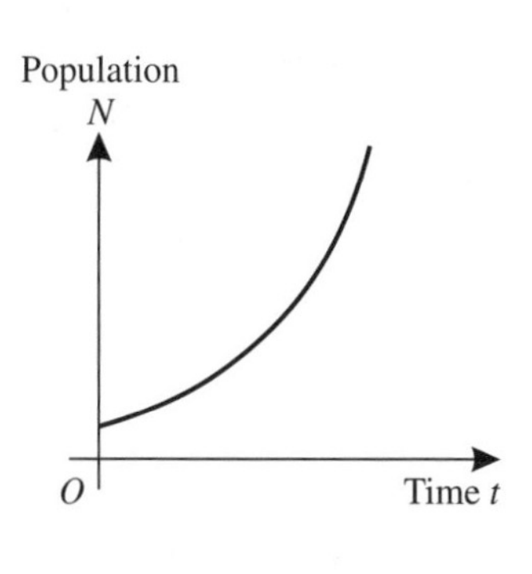

Malthus's model predicts an unlimited growth in population, which is unlikely to occur. In 1837 Pierre Verhulst refined this model by looking at factors which had been ignored in it – in particular factors that might limit population growth. He proposed that the rate of change of a population N is:

$$kN\left(1 - \frac{N}{M}\right)$$

This is based on the assumption that there is an upper limit M to the size of population which can be sustained. The mathematical solution of this problem produces the graph shown here.

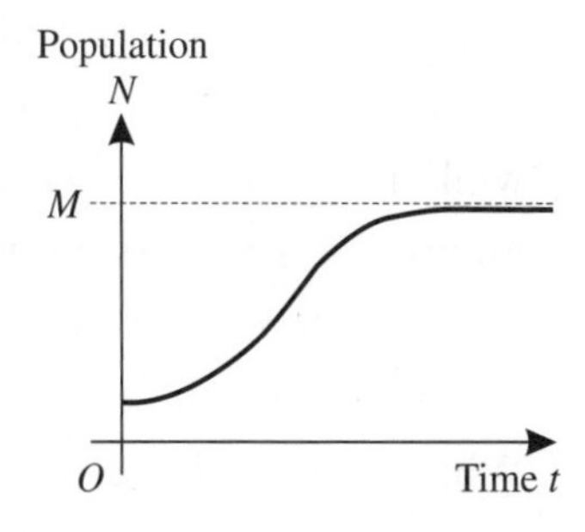

This prediction fitted the population data for the USA from 1810 to 1930. However, it does not agree with population data since then, so further refinements to the model are required.

Exercise 1A

Briefly discuss how you would set up a mathematical model to solve each of the following problems. Include details of which quantities are relevant, any assumptions you have made and any refinements you could make to improve your model.

1 A family of four with two adults and two children aged 8 and 10 wishes to travel to London for a sporting event and to keep their costs to a minimum. Should they travel by car, by coach, by train or some combination of these?

2 A family needs to estimate its household bills for the next three months to decide if it can afford a holiday. Do this for a family like yours.

3 A lecturer has a choice of three routes to drive from college to home. Which one should he choose in rush hour to keep his journey time as short as possible?

4 A woman uses her car for business. She needs to estimate the annual cost of running her car. How can she do this?

5 A company is organising a one-day conference which involves certain expenses. Last year 100 people attended. What conference fee should the company charge each delegate so that it covers its expenses?

6 A new supermarket has opened in a town 10 miles away from where the Browns live. The prices appear to be cheaper than those of their local corner shop. Help the Browns decide whether it will be cheaper to shop locally or at the new supermarket.

7 A family of four adults decides to go to the Channel Islands for a holiday. They agree that they need the use of a car while they are there. They wish to keep the cost to a minimum. The possibilities are (a) fly and hire a car (b) go by boat and hire a car or take their own (c) go on the new catamaran and hire a car or take their own.
Which option should they choose?

1.2 Mathematical models in mechanics

Mathematical modelling involves simplifying a real world problem to produce a model which can be solved mathematically. In mechanics a range of mathematical models has been developed over the years to describe the effects of forces on various types of objects in practical situations.

The basic work on mechanics in this book deals with simple objects and only the most important forces acting on them, allowing you to produce mathematical models which you can solve.

Basic terminology for mechanics

This section introduces the essential terminology you need to know for modelling in mechanics.

- A **particle** is a body whose dimensions are so small compared with the other lengths involved that its position in space can be represented by a single point. For example, in considering the motion of the Earth relative to the Sun you may represent the Earth and the Sun by particles.
- A **bead** is a particle which is assumed to have a hole drilled through it so that it may be threaded onto a string or wire. For example the plastic animal threaded on a wire as part of a child's rattle may be modelled as a bead.
- A **lamina** is a flat object whose thickness is small compared with its width and length. For example, a piece of card, a sheet of paper or a thin metal sheet may be represented by laminae (plural of lamina).
- A **uniform lamina** is one in which equal areas of the lamina have equal masses. This is clearly the case when the whole of the lamina is made of the same material.
- A **rigid body** is an object made up of particles, all of which remain at the same fixed distances from one another whether the object is at rest or in motion. For example, a rigid beam is assumed to keep its shape when acted on by forces. When a billiard ball strikes the edge of a snooker table you may assume that it does not change its shape, so it can be represented by a rigid body.
- A **wire** is a rigid body in the form of a thin thread of metal. The wire may be smooth or rough. We often consider beads threaded on wires.
- A **rod** is an object all of whose mass is concentrated along a line. It is assumed to have length only, and its width and breadth are neglected. A broom pole may, for example, be modelled by a rod.
- A **uniform rod** is one in which equal lengths have equal masses.
- A rod which is not uniform is said to be **non-uniform**. In this case equal lengths do not have equal masses. An example of a non-uniform rod is a baton made in two sections which are different woods.
- A **light object** is one whose mass is so small compared with other masses being considered that the mass may be considered to be zero. A **light string** is one example. If an object is suspended by a light string, the mass of the string may be ignored. A **light rod** is another example. If two particles are joined by a light rod these particles remain the same distance apart. The mass of the rod may be neglected.

- An **inextensible string** or **inelastic string** is a string whose length remains the same whether motion is taking place or not. While all real strings are elastic to some extent, in many problems the extension is so small compared to the other lengths under consideration that it may be ignored.
- A **smooth surface** is one which offers so little frictional resistance to the motion of a body sliding across it that the friction may be ignored. A sheet of ice is an example of a smooth surface commonly found in the real world.
- A surface which is not smooth is said to be **rough**. Frictional forces have to be taken into account when a body moves on such a surface. An example of a rough surface is the surface of a hard tennis court.
- A **smooth pulley** is one with no friction in its bearings.
- A **peg** is a pin or support from which a body may be hung or on which a body may rest. There is only one point of contact between the peg and the body in either case (see figures). The peg may be either smooth or rough. In the first case there is only a contact force. In the second case frictional forces will also act.

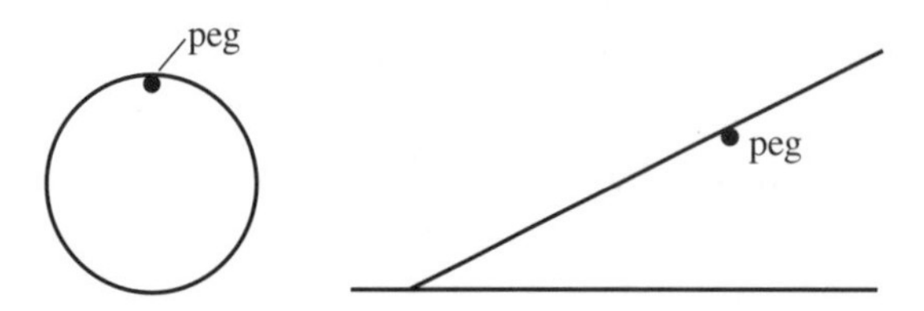

- A **plane surface** is a completely flat surface. Walls, floors and tables are usually modelled by plane surfaces.
- The **Earth's surface** is usually modelled by a horizontal plane surface. (You may also model slopes by inclined planes.)

An example of modelling in mechanics

As an example of modelling in mechanics, think about the motion of a cricket ball thrown through the air. In the simplest possible model the following assumptions are made:

The ball may be modelled by a particle (1)
The acceleration due to gravity is constant (2)
The motion of the ball takes place in a vertical plane (3)
The only force acting on the ball is its weight. (4)

Assumption (1) ignores the size and shape of the ball – sometimes called the **particle model**.

Assumption (2) ignores the variation of gravity with position on the Earth's surface and height above sea level.

Assumption (3) ignores any sideways movement due to crosswinds or spin.

Assumption (4) ignores the effect of any air resistance.

How well does this model match the real world? In the model the path of the ball is a symmetrical curve called a parabola shown here:

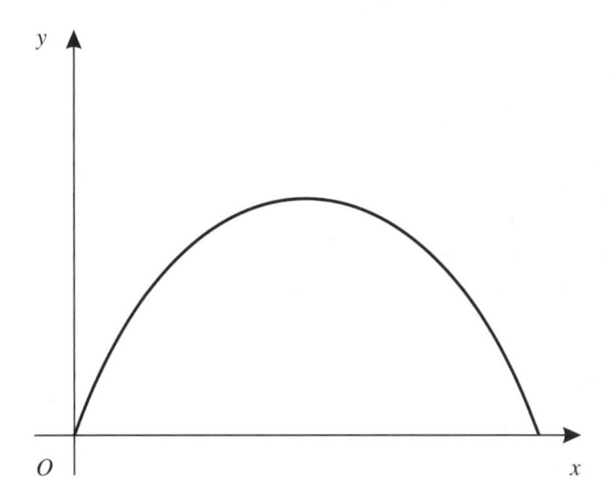

However in the real world the ball's path looks like this:

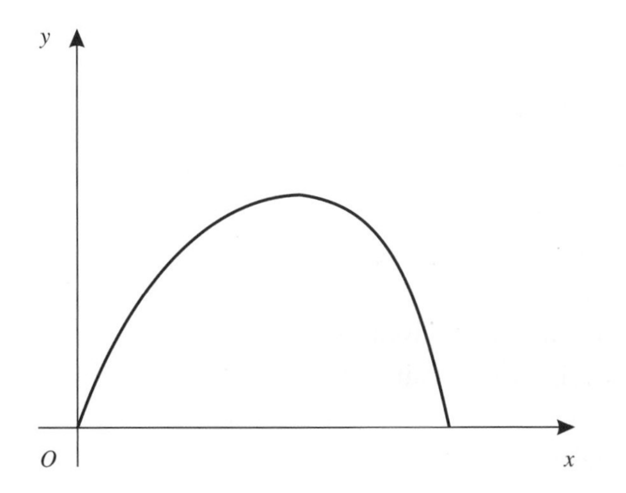

To get a better match to the real world the model needs to be refined. If account is taken of air resistance, which depends on the speed of the ball, the new model is a better match to the real world situation.

Exercise 1B

1 Suggest a simple mathematical model for each of the following problems.

(a) A brick of mass 1 kg is attached to one end of a rope. A boy picks up the other end of the rope and swings the brick so that its path of motion is a vertical circle. Find how the tension in the rope is related to the speed of the brick.

(b) A book of mass 0.5 kg is released from rest on the lid of a polished wood desk inclined at 20° to the horizontal. Find the distance travelled by the book in 5 seconds.
(c) In a game of snooker the white ball strikes a red ball which is at rest. Find how the speed of the red ball, after the collision, is related to the speed of the white ball which strikes it.
(d) A see-saw is made up of a plank of length 4 m balanced on a support at its centre. Andrew has a mass of 25 kg and his sister Brenda has a mass of 20 kg. Andrew sits at one end of the plank. Where should Brenda sit if the see-saw is to balance?
(e) A man needs to climb to the top of a ladder to carry out some repairs. The ladder rests on a rough plane against a rough wall. If you know the man's weight how could you find the maximum inclination of the ladder for it still to be safe to climb?

2 In each of the following a physical situation is given together with a mathematical model. Decide if the model is a good one and suggest ways it could be refined.
(a) A cricket ball is thrown vertically upwards. Given its initial speed find how long it takes to return to its initial position.
Model: A particle moves vertically under constant gravity with no air resistance.
(b) In a game of ice hockey a puck is hit across an ice rink. Given the initial speed find how far it travels in 6 seconds.
Model: A particle moves on a smooth horizontal surface.
(c) A piece of equipment used in a gym consists of two metal spheres each of mass 20 kg and radius 0.1 m joined by a metal rod of mass 0.5 kg and length 2 m. It is dropped from a height of 2 m with the rod horizontal. Determine the speed with which the equipment reaches the ground.
Model: Two particles of mass 20 kg are joined by a light rod of length 2.2 m. The system starts from rest with the rod horizontal and falls freely under gravity.

(d) A builder attaches a large bucket to one end of a rope which passes over a pulley and he uses this to hoist materials to a second floor window. Find the maximum weight the bucket will carry.
Model: A light inextensible string passes over a smooth pulley with a particle attached at the end of the string. Find the tension in the string for a given mass.
(e) In pole-vaulting an athlete uses a glass fibre pole to attempt to clear a horizontal bar. How does his initial speed affect the height he can clear?
Model: A particle of mass equal to that of the athlete is attached at the end of a light rigid rod. The particle moves freely under gravity on leaving the rod.

Vectors and their application in mechanics

2

2.1 Vector and scalar quantities

When applying mathematics to mechanical systems you will often need to carry out calculations involving quantities such as:

- the mass of a particle (for example 10 g)
- the distances between two points on a table (for example 0.5 m)
- the time taken by a runner to complete a 100 m race (say 10 s).

Quantities like these are called **scalars**. Each one is completely described by a number and the appropriate unit of measurement.

It is not too difficult, however, to think of quantities which are *not* completely characterised by a single number.

Suppose you are told that a particle has moved a distance of 5 m from a point O. You do not have enough information to find the final position of the particle. As well as the distance it has moved you need to know the *direction* in which the particle has moved. If the final position of the particle is the point A, then you can say that the **displacement** of the particle is OA. To specify this displacement you need the **length** of OA and the **direction** of $\overrightarrow{OA}$. The arrow shows that the particle has moved from O to A.

Suppose you now are told that a force of a certain size is applied to a particle by means of a string attached to the particle. Again, this information is not sufficient for you to be able to say what happens to the particle. In addition to knowing the **magnitude** or size of the force, you need to know the **direction** in which this force acts.

In both of these examples, to specify the quantities completely you needed to give a number and also a direction.

■ **A vector is defined to be any quantity which is completely specified by a scalar magnitude and a direction.**

Displacements and forces are therefore **vectors**. Later in this chapter you will see that velocities and accelerations are also vectors.

Example 1

A man starts at the point O and walks 4 km east to the point A and then 3 km north to the point B. Find his final displacement from O.

His journey can be shown by drawing a diagram. This is an important first step in a problem of this type.
Using Pythagoras' Theorem:

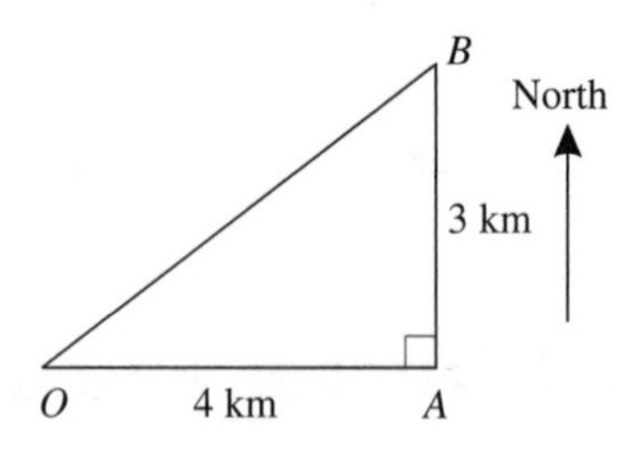

$$\begin{aligned} OB^2 &= OA^2 + AB^2 \\ &= 4^2 + 3^2 \\ &= 16 + 9 = 25 \\ OB &= 5 \end{aligned}$$

The length OB is 5 km.

The *displacement* of the man is given by OB which is of length 5 km.

The *distance* he walked is given by:

$$OA + AB = 7\text{ km}$$

The displacement is represented by the directed straight line $\overrightarrow{OB}$.

Example 2

A man walks 2 km due east from O to A and then 3 km in a north-east direction from A to B. Find the distance of B from O and describe the displacement OB.

First draw a diagram representing the man's journey.

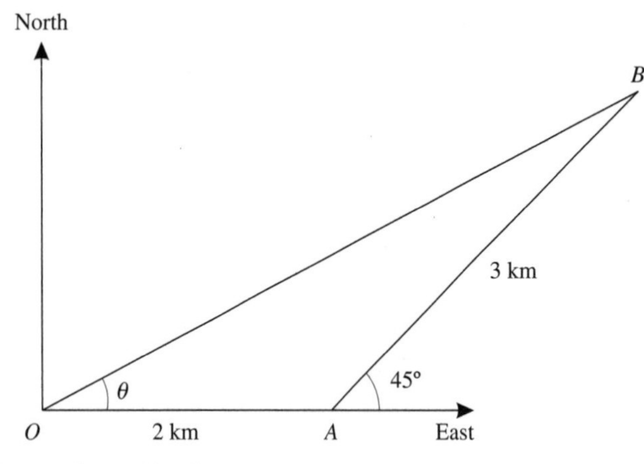

Using the cosine rule gives:

$$OB^2 = OA^2 + AB^2 - 2 \times OA \times AB \cos B\hat{A}O$$

As: $\angle BAO = (180 - 45)^\circ = 135^\circ$

$$OB^2 = 2^2 + 3^2 - 2 \times 2 \times 3 \cos 135^\circ$$

So: $OB = 4.64\text{ km}$

To find the direction of the displacement OB you need to find $\angle BOA$.

Let $\angle BOA = \theta$. Using the sine rule gives:

$$\frac{OB}{\sin 135^\circ} = \frac{AB}{\sin \theta}$$

$$\frac{4.64}{\sin 135^\circ} = \frac{3}{\sin \theta}$$

$$\sin \theta = \frac{3 \sin 135^\circ}{4.64}$$

So: $$\theta = 27.2^\circ$$

Usually when you specify directions you refer them to the north–south line. The angle OB makes with the north line is $(90 - 27.2^\circ) = 62.8^\circ$. So the displacement $\overrightarrow{OB}$ is in the direction N 62.8° E.

Exercise 2A

1 A boy walks x km due east then y km due north. Calculate the total distance he has walked and his displacement from his starting point when:

(a) $x = 5,\ y = 12$ (b) $x = 12,\ y = 5$ (c) $x = 10,\ y = 7$
(d) $x = 1,\ y = 16$ (e) $x = 10,\ y = -27$.

What is the significance of the negative sign in (e)?

2 A girl walks x km due east then z km north-east. Calculate the total distance she has walked and her displacement from her starting point when:

(a) $x = 2,\ z = 5$ (b) $x = 3,\ z = 4$ (c) $x = 5,\ z = 2$
(d) $x = 1,\ z = 6$ (e) $x = 8,\ z = -1$.

3 In a regatta a yacht sails x km due east from a marker O then y km due north to a buoy B. Calculate the magnitude and direction of the displacement vector $\overrightarrow{OB}$ when:

(a) $x = 5,\ y = 8$ (b) $x = 8,\ y = 5$ (c) $x = 10,\ y = 16$.

4 In a desert exercise a tank travels x km on a bearing of 050° from a base O then y km on a bearing of 140° to a bunker B. Calculate the magnitude and direction of the displacement vector $\overrightarrow{OB}$ when:

(a) $x = 7,\ y = 6$ (b) $x = 6,\ y = 7$ (c) $x = 12,\ y = 14$.

5 When orienteering, Jack walks x km on a bearing of 100° from his starting point O then y km on a bearing of 200° to a hut H. Calculate the magnitude and direction of the displacement vector $\overrightarrow{OH}$ when:
(a) $x = 6,\ y = 7$ (b) $x = 6,\ y = 6$ (c) $x = 6,\ y = 10$.

6 Jill is swimming across a river which has a constant current of $k\,\text{m s}^{-1}$. If Jill heads directly across the river with speed $v\,\text{m s}^{-1}$, find her resultant speed and the angle her path makes with a line directly across the river when:
(a) $k = 0.5,\ v = 1$ (b) $k = 0.5,\ v = 1.5$ (c) $k = 2,\ v = 1$.

2.2 Properties of vectors

Representing vectors

Any vector can be represented by a **directed line segment**. The direction of the line segment is that of the vector, and its length represents the magnitude of the vector.

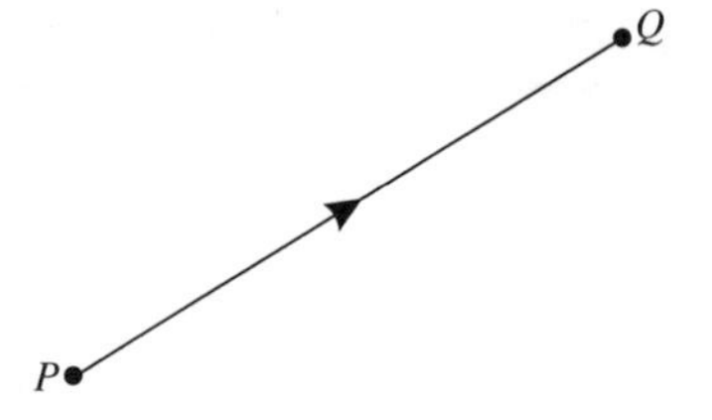

So $\overrightarrow{PQ}$ represents the vector with magnitude and direction given by the line segment joining P and Q.

The magnitude of the vector is written $|\overrightarrow{PQ}|$.

Equal vectors

Two vectors are **equal** if and only if they have both the **same magnitude** and the **same direction**.

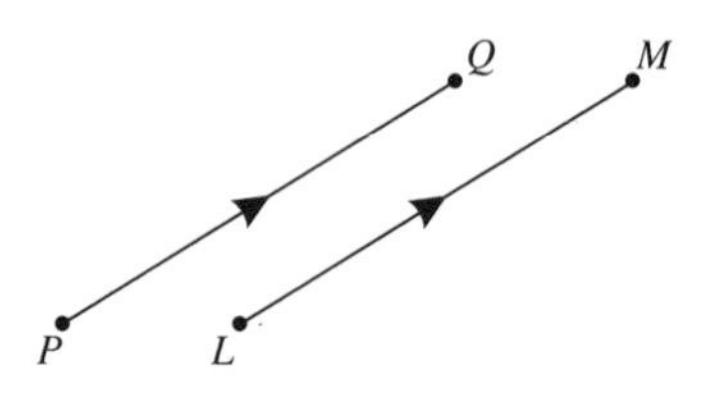

In the diagram, PQ and LM are parallel and also equal in length so $\overrightarrow{PQ} = \overrightarrow{LM}$.

Null vector or zero vector

The magnitude of a vector is usually positive. It is never negative, but it may be zero. In this case it is called the **zero vector** or **null vector**.

Adding vectors

Suppose a runner goes from O to A and then from A to B and a second runner goes directly from O to B.

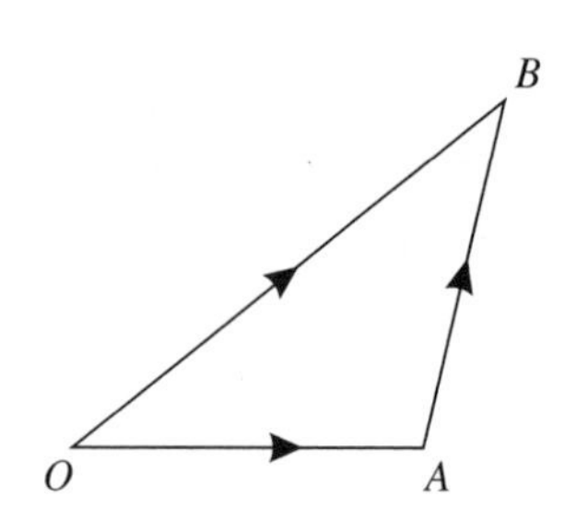

As the result of these two journeys is the same you can write:

$$\overrightarrow{OA} + \overrightarrow{AB} = \overrightarrow{OB}$$

The displacement $\overrightarrow{OB}$ is the **sum** or **resultant** of the displacements $\overrightarrow{OA}$ and $\overrightarrow{AB}$. This rule is called the **triangle law of addition**.

- **All vectors represented by directed line segments can be manipulated in the same way as displacement vectors — and can be added using the triangle law of addition.**

Another way of adding vectors is to use the **parallelogram rule**. If you complete the parallelogram, with sides of $\overrightarrow{OA}$ and $\overrightarrow{AB}$, you get the following:

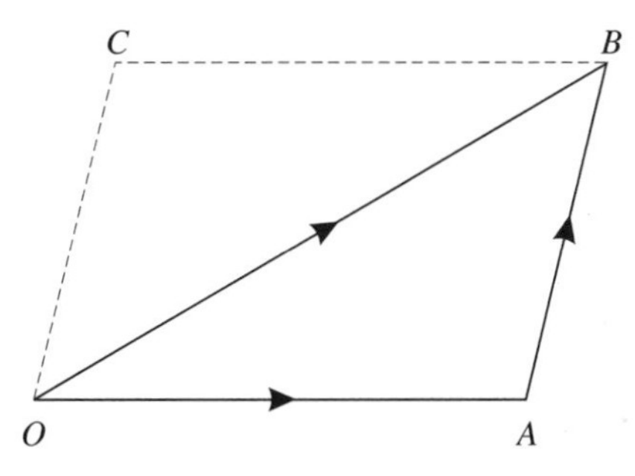

The resultant of $\overrightarrow{OA}$ and $\overrightarrow{AB}$ is then represented by the **diagonal** $\overrightarrow{OB}$ of the parallelogram OABC.

Vectors that are related to each other

Think of a journey from O to A followed by the return journey from A to O. The result of these two journeys is a zero displacement (you end up back where you started). So you can write:

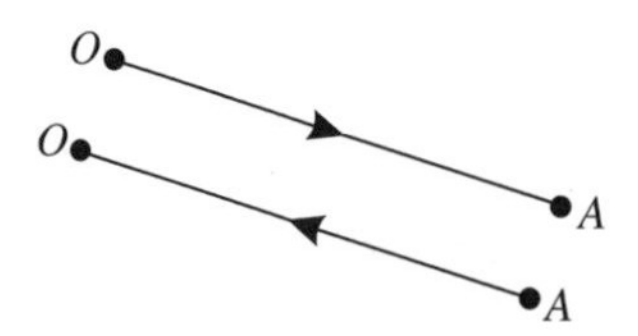

$$\overrightarrow{OA} = -\overrightarrow{AO}$$

You can see that the **order of the letters and the direction of the arrow are important**. $\overrightarrow{OA}$ is not the same as $\overrightarrow{AO}$.

The two displacements $\overrightarrow{PQ}$ and $\overrightarrow{LM}$ are parallel but of different magnitudes.

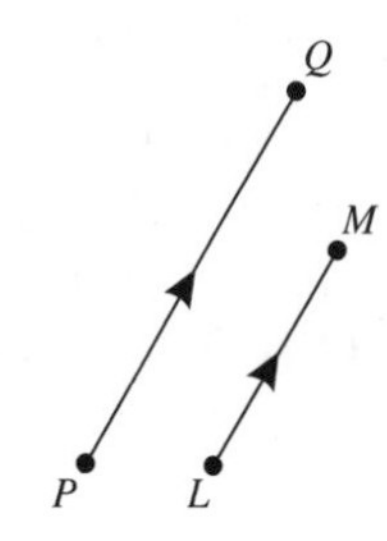

You can write this as:

$$\overrightarrow{PQ} = k\overrightarrow{LM}$$

where k is a scalar that represents the ratio of the magnitudes $|\overrightarrow{PQ}|$ and $|\overrightarrow{LM}|$.

Another type of vector notation

It is often convenient to use an alternative notation in which a vector $\overrightarrow{OA}$ is represented by a single letter, usually written alongside the vector in a diagram. When this is done the letter is always printed in **bold type**. In manuscript it should be written with a wavy line underneath, for example $\underset{\sim}{\mathbf{a}}$. The magnitude of the vector is usually represented by the same letter but printed in *italic*. So:

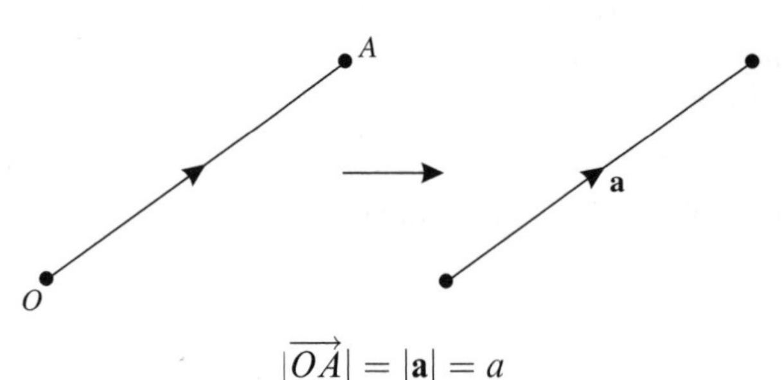

And: $$|\overrightarrow{OA}| = |\mathbf{a}| = a$$

Using this notation the triangle law becomes:

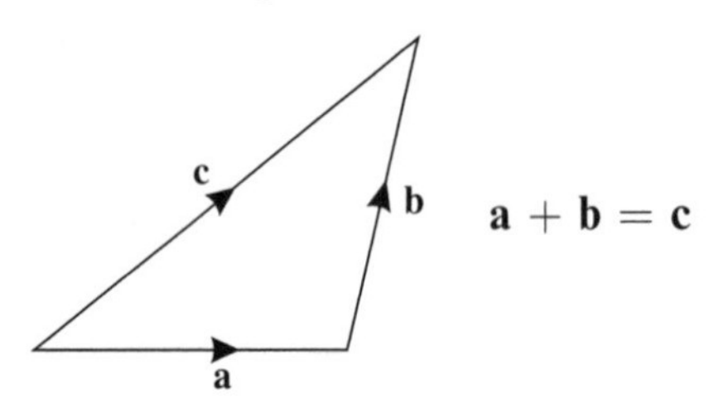

Also, in this notation $-\mathbf{b}$ has the same magnitude as $\mathbf{b}$ but is in the opposite direction:

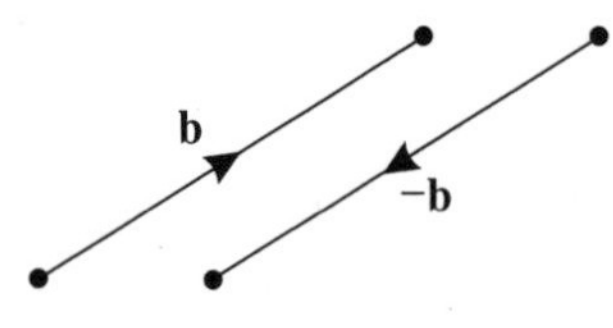

Example 3

The diagram shows a parallelogram $OACB$ with $\overrightarrow{OA} = \mathbf{a}$ and $\overrightarrow{OB} = \mathbf{b}$. The point D is the mid-point of AC. Express the following vectors in terms of **a** and **b**:

(a) $\overrightarrow{BC}$ (b) $\overrightarrow{AC}$ (c) $\overrightarrow{CA}$ (d) $\overrightarrow{OC}$ (e) $\overrightarrow{AB}$ (f) $\overrightarrow{OD}$.

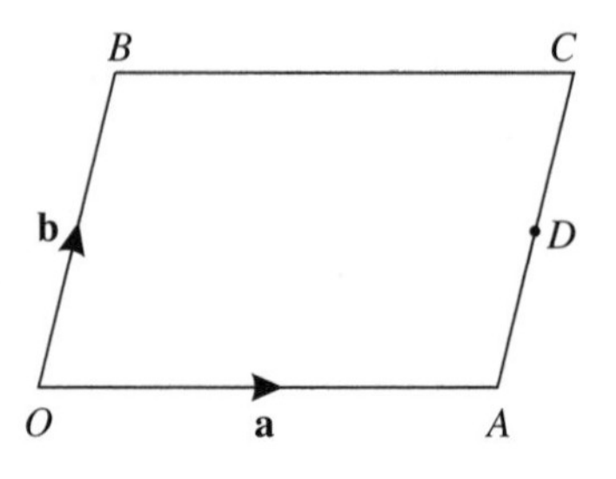

(a) $\overrightarrow{BC}$ is in the same direction as $\overrightarrow{OA}$ and of the same length, because $OACB$ is a parallelogram.

So: $$\overrightarrow{BC} = \overrightarrow{OA} = \mathbf{a}$$

(b) Similarly, $\overrightarrow{AC}$ is in the same direction as $\overrightarrow{OB}$ and of the same length.

So $$\overrightarrow{AC} = \overrightarrow{OB} = \mathbf{b}$$

(c) The vector $\overrightarrow{CA}$ is in the opposite direction to $\overrightarrow{AC}$ but of the same length.

So: $$\overrightarrow{CA} = -\overrightarrow{AC} = -\mathbf{b}$$

(d) To find $\overrightarrow{OC}$, consider the triangle OAC.
By the triangle law:

$$\begin{aligned}\overrightarrow{OC} &= \overrightarrow{OA} + \overrightarrow{AC} \\ &= \overrightarrow{OA} + \overrightarrow{OB} \\ &= \mathbf{a} + \mathbf{b}\end{aligned}$$

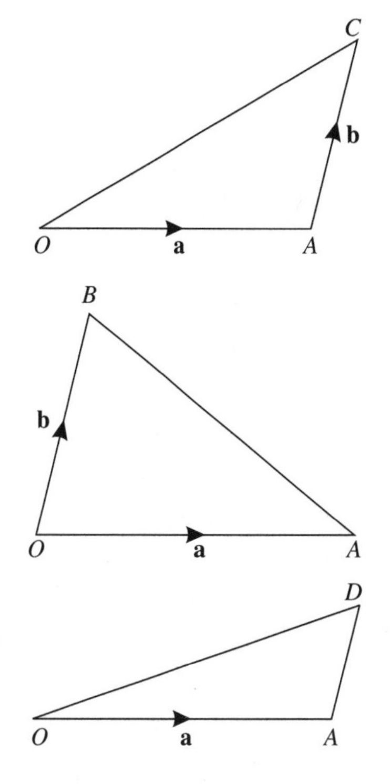

(e) To find $\overrightarrow{AB}$, consider the triangle OAB.
By the triangle law:

$$\overrightarrow{OA} + \overrightarrow{AB} = \overrightarrow{OB}$$

Or:

$$\begin{aligned}\overrightarrow{AB} &= \overrightarrow{OB} - \overrightarrow{OA} \\ &= \mathbf{b} - \mathbf{a}\end{aligned}$$

(f) To find $\overrightarrow{OD}$, consider the triangle OAD.
You must first find $\overrightarrow{AD}$. $\overrightarrow{AD}$ is parallel to $\overrightarrow{AC}$ and half its length as D is the mid-point of $\overrightarrow{AC}$.

So:

$$\overrightarrow{AD} = \tfrac{1}{2}\mathbf{b}$$

By the triangle law:

$$\begin{aligned}\overrightarrow{OD} &= \overrightarrow{OA} + \overrightarrow{AD} \\ &= \mathbf{a} + \tfrac{1}{2}\mathbf{b}\end{aligned}$$

Exercise 2B

1

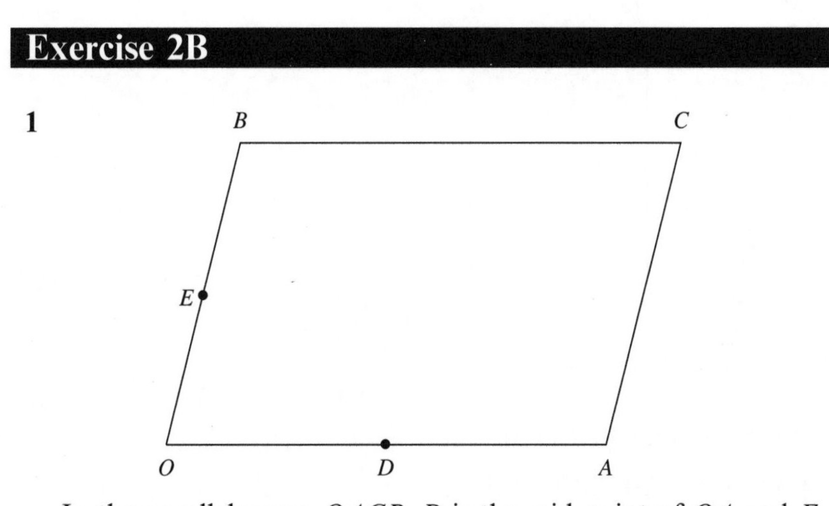

In the parallelogram $OACB$, D is the mid-point of OA and E is the mid-point of OB. Vectors **a** and **b** are represented in magnitude and direction by OA and OB. Find in terms of **a** and **b** the vectors represented in magnitude and direction by:

(a) OD (b) EO (c) AB (d) AC (e) CA
(f) BE (g) DE (h) OC (i) CO (j) DC.

2

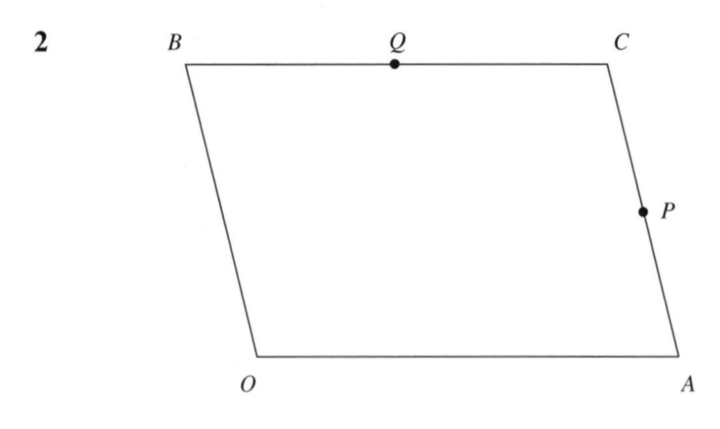

In the parallelogram $OACB$, P is the mid-point of AC and Q is the mid-point of BC. Vectors **a** and **b** are represented in magnitude and direction by OA and OB. Find in terms of **a** and **b** the vectors represented in magnitude and direction by:

(a) AC (b) CB (c) OP (d) AQ
(e) QA (f) QB (g) QP (h) PQ
(i) AB (j) PB.

3

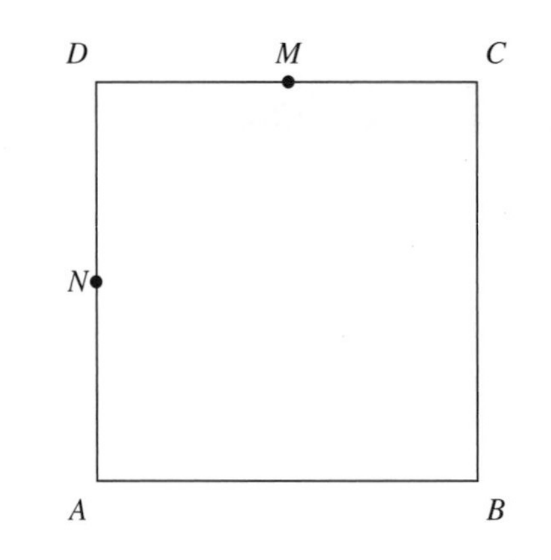

In the square $ABCD$, M is the mid-point of CD and N the mid-point of AD. Vectors **a** and **b** are represented in magnitude and direction by AB and AD. Find in terms of **a** and **b** the vectors represented in magnitude and direction by:

(a) AC (b) BD (c) BM (d) MA
(e) AM (f) NM (g) CN (h) MN
(i) NA (j) CB.

4

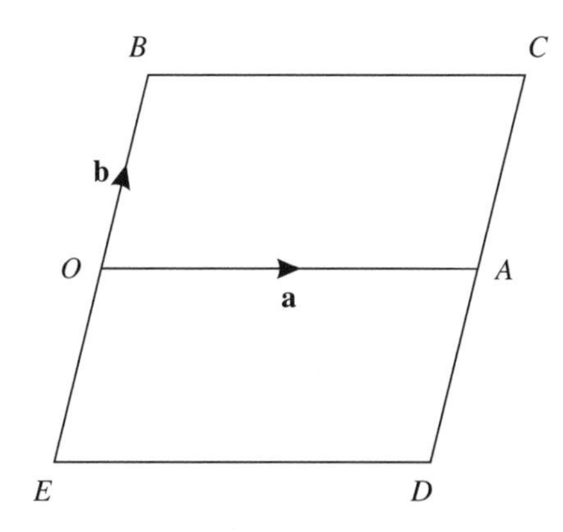

$OACB$ and $OADE$ are two congruent parallelograms. OA and OB represent in magnitude, direction and line of action the vectors **a** and **b**. Find the directed line segments which represent in magnitude, direction and line of action:

(a) $-\mathbf{a}$ (b) $-\mathbf{b}$ (c) $\mathbf{a} + \mathbf{b}$ (d) $\mathbf{a} - \mathbf{b}$ (e) $\mathbf{b} - \mathbf{a}$.

5

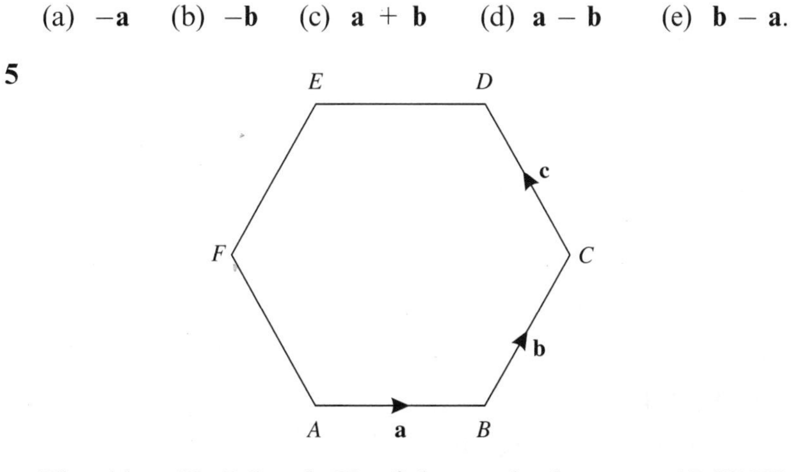

The sides AB, BC and CD of the regular hexagon $ABCDEF$ represent in magnitude and direction the vectors **a**, **b**, and **c** respectively. Find in terms of **a**, **b** and **c** the vectors represented in magnitude and direction by:

(a) DE (b) FE (c) AF (d) AC (e) AD
(f) AE (g) BF (h) EC (i) DF (j) DB.

2.3 Cartesian unit vectors and components

A **unit vector** is a vector with magnitude of 1 unit.
Chapter 5 of Book C1 deals with coordinate geometry. Particularly useful unit vectors are those that lie in the directions of the positive x-axis and the positive y-axis (the cartesian axes) discussed in that chapter.

i, j notation

The unit vectors along the cartesian axes are usually denoted by **i** and **j** respectively:

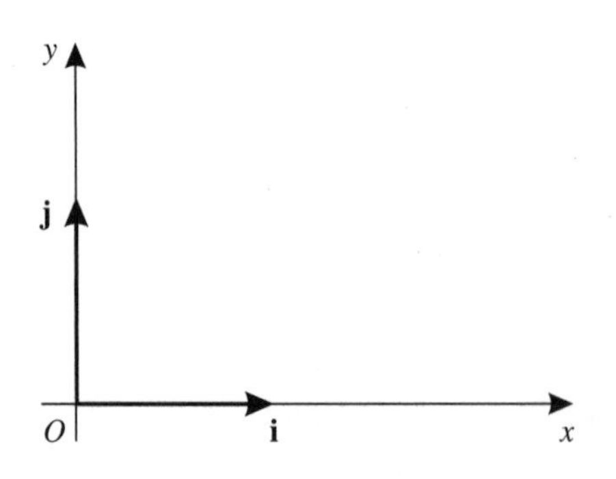

Example 4

A vector (3**i** + 4**j**) units represents a displacement of:

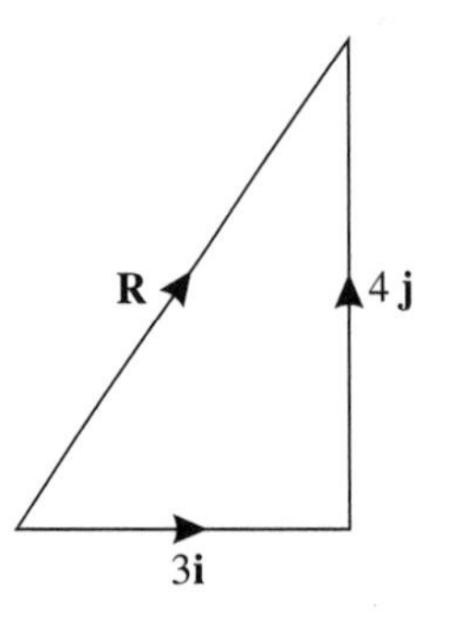

3 units in the direction of the unit vector **i**,
4 units in the direction of the unit vector **j**.

By the triangle law of addition this is just the vector **R** which completes the triangle and so:

$$\mathbf{R} = 3\mathbf{i} + 4\mathbf{j}$$

The vector **R** is written in the **i**, **j** notation. The advantages of the **i**, **j** notation will now be shown.

Adding vectors in i, j notation

When vectors are given in terms of unit vectors **i** and **j**, you can add them together by adding the terms involving **i** and the terms involving **j** separately.

Example 5

Given $\mathbf{a} = 2\mathbf{i} + 3\mathbf{j}$ and $\mathbf{b} = 4\mathbf{i} - 2\mathbf{j}$, find $\mathbf{a} + \mathbf{b}$ in terms of **i** and **j**.

Adding **a** and **b** gives:

$$\begin{aligned}\mathbf{a} + \mathbf{b} &= (2\mathbf{i} + 3\mathbf{j}) + (4\mathbf{i} - 2\mathbf{j}) \\ &= (2\mathbf{i} + 4\mathbf{i}) + (3\mathbf{j} - 2\mathbf{j}) \\ &= 6\mathbf{i} + \mathbf{j}\end{aligned}$$

This is shown in the diagram:

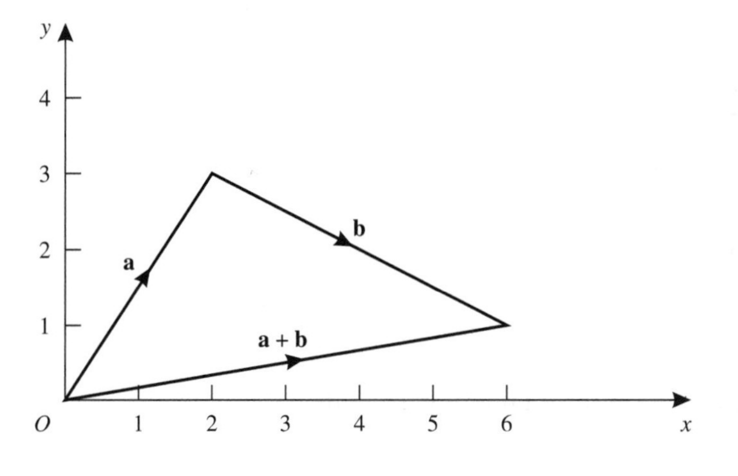

Subtracting vectors in i, j notation

You can subtract vectors given in the **i**, **j** notation in a similar way, by subtracting the **i** parts and **j** parts separately.

Example 6

Given $\mathbf{a} = 6\mathbf{i} + 4\mathbf{j}$ and $\mathbf{b} = 4\mathbf{i} - 3\mathbf{j}$, find $\mathbf{a} - \mathbf{b}$ in terms of **i** and **j**.

Subtracting **b** from **a** gives:

$$\begin{aligned}\mathbf{a} - \mathbf{b} &= (6\mathbf{i} + 4\mathbf{j}) - (4\mathbf{i} - 3\mathbf{j}) \\ &= (6\mathbf{i} - 4\mathbf{i}) + (4\mathbf{j} - (-)3\mathbf{j}) \\ &= 2\mathbf{i} + 7\mathbf{j}\end{aligned}$$

Magnitude of a vector in i, j notation

When a vector **R** is given in terms of the unit vectors **i** and **j** you can find its magnitude by using Pythagoras' Theorem. You can show that $\mathbf{R} = x\mathbf{i} + y\mathbf{j}$ as:

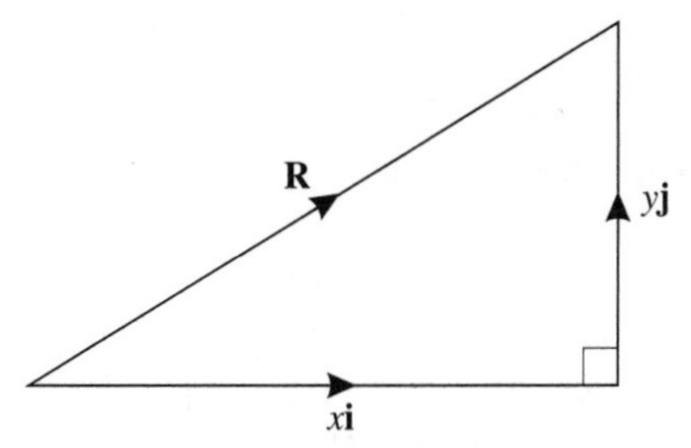

By Pythagoras' Theorem, $R^2 = x^2 + y^2$ and so the magnitude of $x\mathbf{i} + y\mathbf{j}$ is $\sqrt{(x^2 + y^2)}$.

Example 7

Find the unit vector in the direction of the vector $\mathbf{a} = 2\mathbf{i} + 3\mathbf{j}$.

The magnitude of $\mathbf{a}$ is:

$$|\mathbf{a}| = \sqrt{(2^2 + 3^2)} = \sqrt{13}$$

So the *unit* vector in the same direction as $\mathbf{a}$ is:

$$\frac{\mathbf{a}}{|\mathbf{a}|} = \frac{1}{\sqrt{13}}(2\mathbf{i} + 3\mathbf{j})$$

Equality of vectors

Suppose the two vectors $\mathbf{a}$ and $\mathbf{b}$ are given in the $\mathbf{i}$, $\mathbf{j}$ notation so that

$$\mathbf{a} = a_1\mathbf{i} + a_2\mathbf{j}$$
$$\mathbf{b} = b_1\mathbf{i} + b_2\mathbf{j}$$

These two vectors are equal only if:

$$a_1 = b_1 \text{ and } a_2 = b_2$$

On the other hand, if $a_1 = b_1$ and $a_2 = b_2$, then:

$$\mathbf{a} = \mathbf{b}$$

Example 8

Two vectors $\mathbf{a}$ and $\mathbf{b}$ are given by:

$$\mathbf{a} = (2 - x)\mathbf{i} + 4\mathbf{j}$$
$$\mathbf{b} = \mathbf{i} + (6 - y)\mathbf{j}$$

where x and y are constants. Find x and y such that $\mathbf{a} = \mathbf{b}$.

If $\mathbf{a} = \mathbf{b}$ then: $(2 - x)\mathbf{i} + 4\mathbf{j} = \mathbf{i} + (6 - y)\mathbf{j}$

Equating the $\mathbf{i}$ parts gives: $2 - x = 1$

So: $x = 1$

Equating the $\mathbf{j}$ parts gives: $4 = 6 - y$

So: $y = 2$

Components of a vector

Any two vectors can be combined, using the triangle law of addition, into a single resultant (see p.15). It is possible and often useful to reverse this process; that is, to replace a vector by an equivalent pair of vectors. This process is called **resolving a vector into components**. This is very important in statics and will be considered further in chapter 4 with reference to forces. At this point we will consider only component vectors in the direction of the coordinate axes.

Look at the situation shown in the diagram.

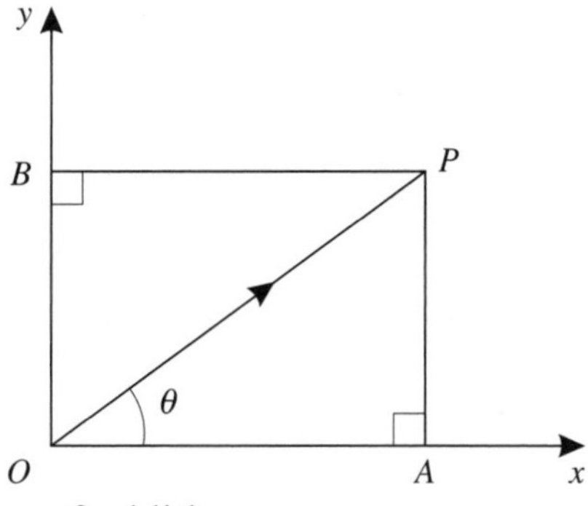

By the triangle law of addition:

$$\overrightarrow{OP} = \overrightarrow{OA} + \overrightarrow{AP}$$

But: $\overrightarrow{AP} = \overrightarrow{OB}$

So: $\overrightarrow{OP} = \overrightarrow{OA} + \overrightarrow{OB}$

By trigonometry: $\dfrac{OA}{OP} = \cos\theta \Rightarrow OA = OP\cos\theta$

$$\frac{OB}{OP} = \sin\theta \Rightarrow OB = OP\sin\theta$$

So: $\overrightarrow{OA} = OP\cos\theta\,\mathbf{i}$

$$\overrightarrow{OB} = OP\sin\theta\,\mathbf{j}$$

giving: $\overrightarrow{OP} = (OP\cos\theta)\,\mathbf{i} + (OP\sin\theta)\,\mathbf{j}$

$OP\cos\theta$ and $OP\sin\theta$ are called the **components**, or **resolutes,** of $\overrightarrow{OP}$ along the coordinate axes. (They are sometimes called the **cartesian components**.)

Example 9

A displacement $\mathbf{R}$ is of magnitude of 16 km on a bearing of 030°. Take $\mathbf{i}$ as a unit vector due east and $\mathbf{j}$ as a unit vector due north and write $\mathbf{R}$ in the form $a\mathbf{i} + b\mathbf{j}$.

This information can be represented by a diagram:

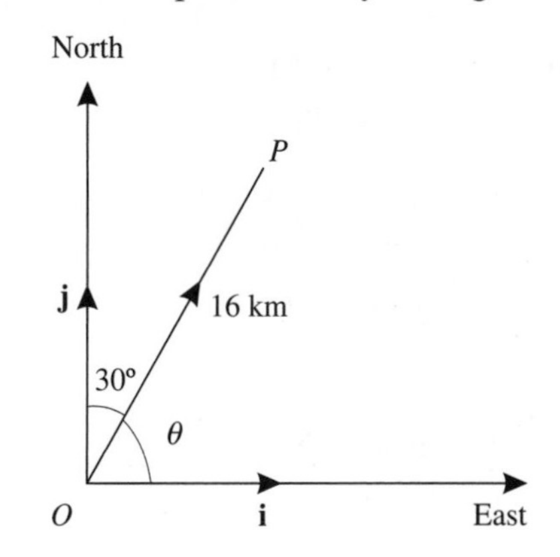

The angle θ is $90° - 30° = 60°$.

So: $$\mathbf{R} = (R\cos\theta)\,\mathbf{i} + (R\sin\theta)\,\mathbf{j}$$
$$= 16\cos 60°\,\mathbf{i} + 16\sin 60°\,\mathbf{j}$$

and: $$\mathbf{R} = 8(\mathbf{i} + \sqrt{3}\mathbf{j})\ \text{km}$$

using $\cos 60° = \frac{1}{2}$ and $\sin 60° = \frac{\sqrt{3}}{2}$, (see chapter 2 of Book P1).

Example 10

The vector **V** is in the direction of the vector $8\mathbf{i} - 6\mathbf{j}$ and has magnitude 5. Find **V** in the form $a\mathbf{i} + b\mathbf{j}$.

V is in the direction of $(8\mathbf{i} - 6\mathbf{j})$, so:

$$\mathbf{V} = k(8\mathbf{i} - 6\mathbf{j})$$

where k is a positive constant that you must find.

As: $$\mathbf{V} = 8k\mathbf{i} - 6k\mathbf{j}$$
$$|\mathbf{V}| = \sqrt{(64k^2 + 36k^2)} = 10k$$

But **V** has magnitude 5 and so:

$$10k = 5$$
$$k = \tfrac{1}{2}$$

Hence: $$\mathbf{V} = \tfrac{1}{2}(8\mathbf{i} - 6\mathbf{j})$$
$$= 4\mathbf{i} - 3\mathbf{j}$$

In a case like this it is worth checking that the vector you have obtained *does* have the required magnitude.

Exercise 2C

In these questions, **i** and **j** are unit vectors in the direction of the positive x- and y-axes respectively.

1 Given that $\mathbf{a} = \mathbf{i} + 2\mathbf{j}$ and $\mathbf{b} = 2\mathbf{i} - \mathbf{j}$ find in terms of **i** and **j**:
(a) $\mathbf{a} + \mathbf{b}$ (b) $\mathbf{a} + 2\mathbf{b}$ (c) $2\mathbf{a} + \mathbf{b}$ (d) $\mathbf{a} - 2\mathbf{b}$
(e) $2\mathbf{a} - \mathbf{b}$.

2 Given that $\mathbf{a} = 2\mathbf{i} + \mathbf{j}$ and $\mathbf{b} = \mathbf{i} + 3\mathbf{j}$ find (a) λ, if $\mathbf{a} + \lambda\mathbf{b}$ is parallel to the vector **i** (b) μ, if $\mu\mathbf{a} + \mathbf{b}$ is parallel to the vector **j**.

3 Given that $\mathbf{a} = \mathbf{i} - 2\mathbf{j}$ and $\mathbf{b} = -3\mathbf{i} + \mathbf{j}$ find (a) λ, if $\mathbf{a} + \lambda\mathbf{b}$ is parallel to $-\mathbf{i} - 3\mathbf{j}$ (b) μ, if $\mu\mathbf{a} + \mathbf{b}$ is parallel to $2\mathbf{i} + \mathbf{j}$.

4 (a) Calculate the magnitude of each of the following vectors:
(i) $5\mathbf{i} + 12\mathbf{j}$ (ii) $5\mathbf{i} + 7\mathbf{j}$ (iii) $5\mathbf{i} - 7\mathbf{j}$ (iv) $-5\mathbf{i} - 8\mathbf{j}$
(v) $7\mathbf{i} - 10\mathbf{j}$.
(b) Calculate the angle made with the positive x-axis by each of the five vectors above.

5 Given that $\mathbf{a} = 2\mathbf{i} + 3\mathbf{j}$ find the magnitude of each of the following vectors and the angle each makes with Ox.
(a) $\mathbf{a}$ (b) $-\mathbf{a}$ (c) $2\mathbf{a}$ (d) $\frac{1}{2}\mathbf{a}$ (e) $-5\mathbf{a}$.

6 Given that $\mathbf{a} = \mathbf{i} + \mathbf{j}$ and $\mathbf{b} = 2\mathbf{i} - \mathbf{j}$ find the magnitude of each of the following vectors and the angle each makes with $0x$.
(a) $\mathbf{a} + \mathbf{b}$ (b) $\mathbf{a} - \mathbf{b}$ (c) $2\mathbf{a} + \mathbf{b}$ (d) $3\mathbf{a} - 2\mathbf{b}$
(e) $3\mathbf{b} - \mathbf{a}$.

7 Find a unit vector in the direction of each of the following vectors:
(a) $\mathbf{i} + \mathbf{j}$ (b) $\mathbf{i} - \mathbf{j}$ (c) $3\mathbf{i} - 4\mathbf{j}$ (d) $-3\mathbf{i} + 4\mathbf{j}$
(e) $12\mathbf{i} + 16\mathbf{j}$.

8

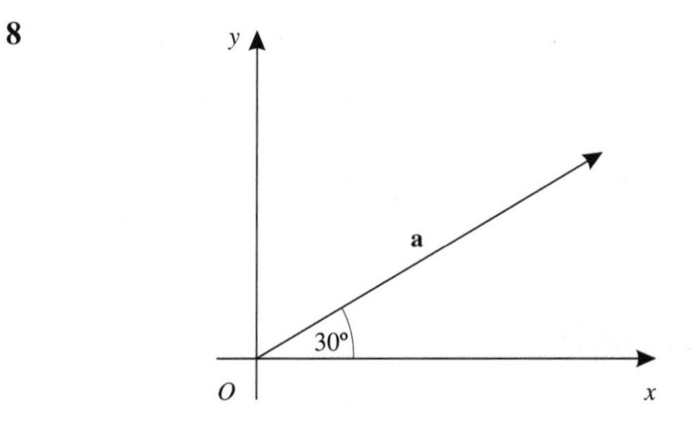

The vector $\mathbf{a}$ is in the plane Oxy, where Ox, Oy are the coordinate axes, and makes an angle 30° with Ox, as in the above diagram. Find the components of $\mathbf{a}$ along Ox and Oy when a equals:
(a) 2 (b) 3 (c) 6 (d) 10.

9 The vector $\mathbf{a}$ is in the plane of the coordinate axes Ox and Oy as in question 8, but makes an angle θ with the positive direction Ox. Find the components of $\mathbf{a}$ along Ox and Oy when $a = 10$ and θ equals:
(a) 60° (b) 120° (c) 150° (d) 240° (e) 300°.

10 Vectors **a** and **b** are in the plane of the coordinate axes Ox and Oy, **a** makes an angle α with Ox, **b** makes an angle β with Ox. Find:

(a) the components of **a** and **b** along Ox, Oy,

(b) the sum of these components along Ox, Oy,

(c) the magnitude of $\mathbf{a} + \mathbf{b}$,

in each of the following cases.

(i) $a = 3,\ b = 4,\ \alpha = 0,\ \beta = 90°$

(ii) $a = 3,\ b = 4,\ \alpha = 0,\ \beta = 60°$

(iii) $a = 3,\ b = 4,\ \alpha = 30°,\ \beta = 60°$

(iv) $a = 5,\ b = 8,\ \alpha = 30°,\ \beta = 60°$

(v) $a = 5,\ b = 8,\ \alpha = 30°,\ \beta = 150°$.

2.4 Using vectors in mechanics

Position vectors

Imagine a particle P moving in a plane. O is a fixed point in the plane. If you know where the point O is, then the position of P is uniquely defined by the vector:

$$\overrightarrow{OP} = \mathbf{r}$$

The vector **r** is called the **position vector** of P relative to O.

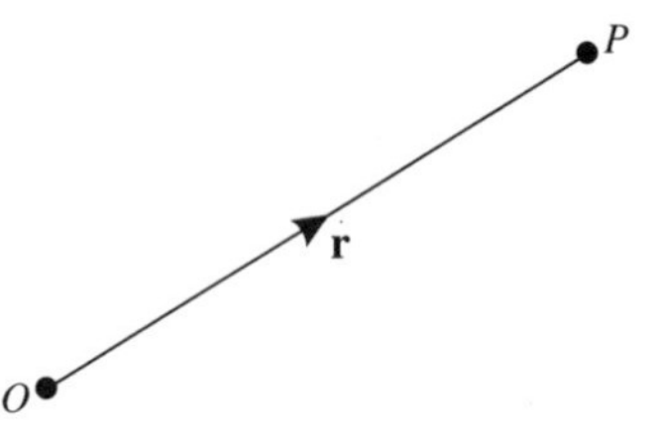

Example 11

At a given time the cartesian coordinates of the position P of a particle are (2,1). Find the position vector of P relative to O.

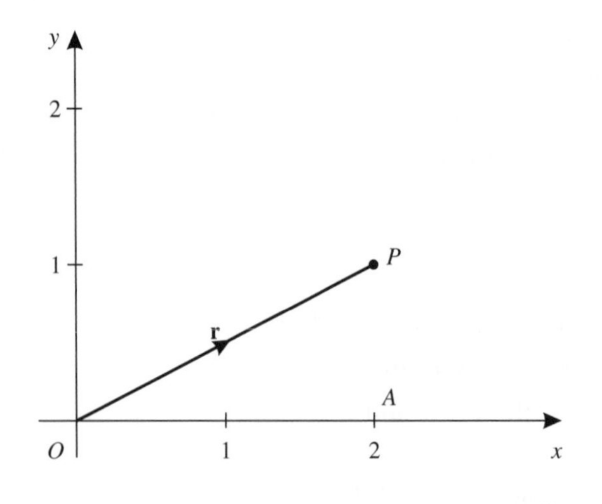

The diagram shows the position of P. Using the **i**, **j** notation:

$$\mathbf{r} = \overrightarrow{OP} = \overrightarrow{OA} + \overrightarrow{AP}$$
$$= 2\mathbf{i} + \mathbf{j}$$

Relative position vector

Imagine two particles C and D moving in a plane; O is a fixed point in the plane. Then $\overrightarrow{OC} = \mathbf{r}_C$ is the position vector of C relative to O, and $\overrightarrow{OD} = \mathbf{r}_D$ is the position vector of D relative to O.

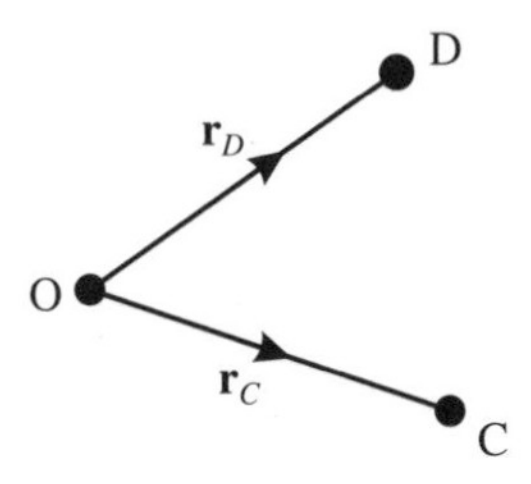

The vector $\overrightarrow{CD}$ gives the position vector of D relative to C. It is called the **relative position vector**.

From the triangle law of addition

$$\overrightarrow{OD} = \overrightarrow{OC} + \overrightarrow{CD}$$

or
$$\overrightarrow{CD} = \overrightarrow{OD} - \overrightarrow{OC} = \mathbf{r}_D - \mathbf{r}_C$$

- **The position vector of D relative to $C = \mathbf{r}_D - \mathbf{r}_C$**

The position vector of C relative to D is:

$$\overrightarrow{DC} = -\overrightarrow{CD} = \mathbf{r}_C - \mathbf{r}_D$$

the negative of the position vector of D relative to C.

Example 12

At a given time particle C has position vector $(4\mathbf{i} - 6\mathbf{j})$ m relative to a fixed origin O and particle D has position vector $(3\mathbf{i} + 2\mathbf{j})$ m relative to O. Find the position vector of D relative to C.

The position vector of D relative to C is:

$$\mathbf{r}_D - \mathbf{r}_C = [(3\mathbf{i} + 2\mathbf{j}) - (4\mathbf{i} - 6\mathbf{j})]\text{ m}$$
$$= [-\mathbf{i} + 8\mathbf{j}]\text{ m}$$

Velocity as a vector

If a particle is moving with a *speed* of $5\,\text{m s}^{-1}$ then, provided its speed remains constant, it will travel 5 m in every second. This is true no matter what type of path the particle is moving on. It will travel 10 m in 2 s, 15 m in 3 s and so on. So in this case:

$$\text{distance travelled} = \text{speed} \times \text{time}$$

But if you want a complete picture of what is happening, you also need to know the *direction* in which the particle is moving. The **velocity** of the particle gives us a complete picture.

- **The velocity of a particle is a vector in the direction of motion whose magnitude is equal to the speed of the particle.**

It is usually denoted by $\mathbf{v}$.

(It can be shown that if the position vector of the particle is $\mathbf{r}$ then $\mathbf{v} = \frac{\mathrm{d}\mathbf{r}}{\mathrm{d}t}$ that is, $\mathbf{v}$ is the rate of change of $\mathbf{r}$ with respect to t – see chapter 5 of Book P1.)

When distances are measured in metres and time in seconds, velocities are measured in metres per second (m s^{-1}).

If the velocity is a constant vector then:

$$\text{displacement} = \text{velocity} \times \text{time}$$

Relative velocity

Consider again two particles C and D moving in a plane. If the velocity of C is $\mathbf{v}_C$ and that of D is $\mathbf{v}_D$ then:

- **the velocity of D relative to C is $\mathbf{v}_D - \mathbf{v}_C$.**

Example 13

At a given time, particle C has velocity $(4\mathbf{i} + 3\mathbf{j})\,\text{m s}^{-1}$ and particle D has velocity $(5\mathbf{i} + 2\mathbf{j})\,\text{m s}^{-1}$. Find the velocity of D relative to C.

The velocity of D relative to C is

$$\begin{aligned}\mathbf{v}_D - \mathbf{v}_C &= [(5\mathbf{i} + 2\mathbf{j}) - (4\mathbf{i} + 3\mathbf{j})]\,\text{m s}^{-1} \\ &= [\mathbf{i} - \mathbf{j}]\,\text{m s}^{-1}\end{aligned}$$

Example 14

A particle P is moving along the x-axis with a constant speed. At time $t = 0$ P is at the origin. At time $t = 3\,\text{s}$ the particle is at the point (9,0). Find (a) the speed of P, given that the distances are measured in metres, (b) the velocity of P, taking $\mathbf{i}$ to be the unit vector in the direction of the positive x-axis.

(a) Distance moved in 3 s by P is 9 m.

So: $$\text{speed of } P = \frac{\text{distance travelled}}{\text{time}} = \frac{9}{3} = 3$$

The speed of P is $3\,\text{m}\,\text{s}^{-1}$.

(b) As the distance moved is along the positive x-axis the velocity of P is $3\mathbf{i}\,\text{m}\,\text{s}^{-1}$.

Example 15

A particle P is at the origin O at time $t = 0$. The particle moves with constant velocity and passes through the point with position vector $(6\mathbf{i} + 8\mathbf{j})$ m, relative to O, at time $t = 2$s, where $\mathbf{i}$, $\mathbf{j}$ are unit vectors along the positive x- and y-axes respectively. Find the velocity of P.

From the information given, you see that the displacement of P in 2 s is $(6\mathbf{i} + 8\mathbf{j})$ m.

So: $$\text{velocity} = \frac{\text{displacement}}{\text{time}} = \frac{6\mathbf{i} + 8\mathbf{j}}{2}$$

The velocity of P is $(3\mathbf{i} + 4\mathbf{j})\,\text{m}\,\text{s}^{-1}$.

Acceleration as a vector

Everybody has some idea of acceleration from their experience of travelling in buses or cars. Just as velocity tells you how the position of a particle changes with time, the **acceleration** tells you how the velocity changes with time. Since velocity has magnitude and direction so also has acceleration. It is, therefore, a vector. It is usually denoted by $\mathbf{a}$. (It can be shown that $\mathbf{a} = \frac{\mathrm{d}\mathbf{v}}{\mathrm{d}t}$.)

- **Acceleration is the rate of change of velocity with respect to time.**

Hence: $$\text{acceleration} = \frac{\text{change in velocity}}{\text{time}}$$

when the acceleration is constant.

Since velocities are measured in metres per second ($\text{m}\,\text{s}^{-1}$) the acceleration is measured in metres per second per second ($\text{m}\,\text{s}^{-2}$).

Example 16

A particle P has velocity $(3\mathbf{i} + 2\mathbf{j})\,\text{m}\,\text{s}^{-1}$ when $t = 0$ and velocity $(7\mathbf{i} + 4\mathbf{j})\,\text{m}\,\text{s}^{-1}$ at time $t = 2$ s. The acceleration of P is constant. Find the acceleration.

The change in velocity of P is:

$$\begin{aligned}(7\mathbf{i} + 4\mathbf{j}) - (3\mathbf{i} + 2\mathbf{j}) &= (7\mathbf{i} - 3\mathbf{i}) + (4\mathbf{j} - 2\mathbf{j}) \\ &= (4\mathbf{i} + 2\mathbf{j})\,\text{m}\,\text{s}^{-1}\end{aligned}$$

This takes place in 2 seconds.

$$\text{So acceleration} = \frac{4\mathbf{i} + 2\mathbf{j}}{2}$$

The acceleration of P is $(2\mathbf{i} + \mathbf{j})\,\text{m}\,\text{s}^{-2}$.

Example 17

A particle P is moving with a constant velocity $(12\mathbf{i} + 5\mathbf{j})\,\text{m s}^{-1}$. It passes through the point A whose position vector is $(4\mathbf{i} + 5\mathbf{j})\,\text{m}$ at $t = 0$. Find:

(a) the speed of the particle
(b) the distance of P from O when $t = 3\,\text{s}$.

(a) The speed of the particle is the magnitude of $\mathbf{v}$, where $\mathbf{v} = (12\mathbf{i} + 5\mathbf{j})\,\text{m s}^{-1}$. So:

$$|\mathbf{v}| = \sqrt{(12^2 + 5^2)} = \sqrt{169} = 13$$

The speed of P is $13\,\text{m s}^{-1}$.

(b) Suppose when $t = 3\,\text{s}$ the particle is at the point B as shown in the diagram.

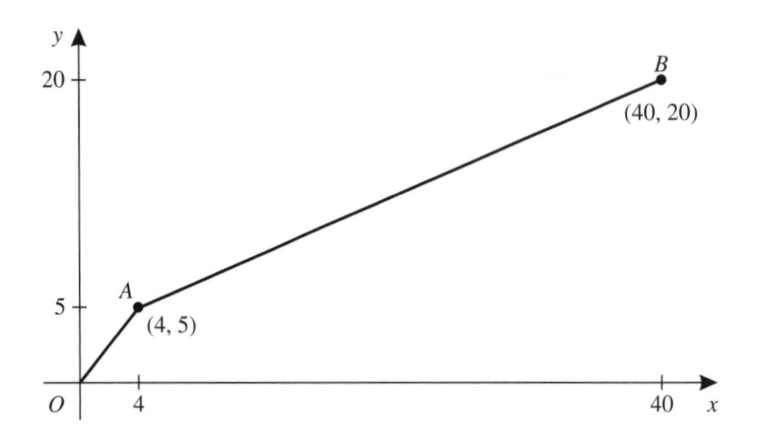

Using: displacement = velocity × time

Gives:

$$\overrightarrow{AB} = (12\mathbf{i} + 5\mathbf{j}) \times 3$$
$$= 36\mathbf{i} + 15\mathbf{j}$$

The position vector of B relative to O is:

$$\overrightarrow{OB} = \overrightarrow{OA} + \overrightarrow{AB}$$
$$= (4\mathbf{i} + 5\mathbf{j}) + (36\mathbf{i} + 15\mathbf{j})$$
$$= 40\mathbf{i} + 20\mathbf{j}$$

So the distance of B from O is:

$$|\overrightarrow{OB}| = \sqrt{(40^2 + 20^2)}$$
$$= \sqrt{2000}$$
$$= 44.7\,\text{m}$$

Example 18

At noon a lighthouse keeper observes two ships A and B which have position vectors $(-4\mathbf{i} + 3\mathbf{j})$ km and $(4\mathbf{i} + 9\mathbf{j})$ km, respectively, relative to the lighthouse O. (The unit vectors $\mathbf{i}$ and $\mathbf{j}$ are directed due east and due north.) The ships are moving with constant velocities $(4\mathbf{i} + 17\mathbf{j})\,\text{km h}^{-1}$ and $(-12\mathbf{i} + 5\mathbf{j})\,\text{km h}^{-1}$ respectively.

(a) Write down in the form $\mathbf{p} + \mathbf{q}t$ the position vector of A and the position vector of B at time t h after noon.
(b) Show that A and B will collide and find the time when this collision will occur and the position vector of the point of collision.

In order to prevent a collision, at 12.15 p.m. ship A increases its speed and changes its velocity to $(16\mathbf{i} + 17\mathbf{j})\,\text{km h}^{-1}$. Ship B maintains its original velocity.

(c) Find the distance between A and B at 12.30 now.

(a) Since the velocity of A is constant its displacement is given by velocity $\times$ time $= (4\mathbf{i} + 17\mathbf{j})\,t$ km. So the position vector $\mathbf{r}_\text{A}$ km of ship A is:

$$(\text{initial position}) + (\text{displacement})$$

$$\mathbf{r}_A = (-4\mathbf{i} + 3\mathbf{j}) + (4\mathbf{i} + 17\mathbf{j})\,t$$

In a similar way, since B has initial position $(4\mathbf{i} + 9\mathbf{j})$ km and constant velocity $(-12\mathbf{i} + 5\mathbf{j})\,\text{km h}^{-1}$ its position vector at time t h is:

$$\mathbf{r}_B = (4\mathbf{i} + 9\mathbf{j}) + (-12\mathbf{i} + 5\mathbf{j})\,t$$

(b) The ships will collide if they have the same position at the same time.
If $\mathbf{r}_A = \mathbf{r}_B$ then:

$$(-4\mathbf{i} + 3\mathbf{j}) + (4\mathbf{i} + 17\mathbf{j})\,t = (4\mathbf{i} + 9\mathbf{j}) + (-12\mathbf{i} + 5\mathbf{j})\,t$$

or $$(-4 + 4t)\,\mathbf{i} + (3 + 17t)\,\mathbf{j} = (4 - 12t)\,\mathbf{i} + (9 + 5t)\,\mathbf{j}$$

Equating the $\mathbf{i}$ parts gives:

$$-4 + 4t = 4 - 12t \qquad \text{(i)}$$

Equating the $\mathbf{j}$ parts gives:

$$3 + 17t = 9 + 5t \qquad \text{(ii)}$$

If the ships collide, then both equation (i) and equation (ii) will be satisfied by *the same value of t*.

From (i): $\quad 4t + 12t = 4 + 4 \Rightarrow t = \frac{1}{2}$

Substituting into (ii): $\quad \text{LHS} = 11\frac{1}{2}$

$$\text{RHS} = 11\tfrac{1}{2} = \text{LHS}$$

So the ships do collide when $t = \frac{1}{2}$, that is at 12.30 p.m. The position vector of the point of collision is:

$$\begin{aligned}\mathbf{r} &= (-4\mathbf{i} + 3\mathbf{j}) + (4\mathbf{i} + 17\mathbf{j})\tfrac{1}{2} \\ &= -2\mathbf{i} + 11\tfrac{1}{2}\,\mathbf{j}\end{aligned}$$

This answer can be checked by substituting $t = \frac{1}{2}$ in $\mathbf{r}_B$.

(c) The position vector of A will now be made up of three parts:

(initial displacement)
+ (displacement from 12 noon to 12.15 p.m.)
+ (displacement from 12.15 p.m. to 12.30 p.m.)

Displacement from 12 noon to 12.15 p.m. $= (4\mathbf{i} + 17\mathbf{j})\frac{1}{4}$

(Remember t is measured in hours.)

Displacement from 12.15 p.m. to 12.30 p.m. $= (16\mathbf{i} + 17\mathbf{j})\frac{1}{4}$

Hence the position vector of A at 12.30 p.m. is now:

$$\begin{aligned}\mathbf{r}_A &= (-4\mathbf{i} + 3\mathbf{j}) + \tfrac{1}{4}(4\mathbf{i} + 17\mathbf{j}) + \tfrac{1}{4}(16\mathbf{i} + 17\mathbf{j}) \\ &= \mathbf{i} + 11\tfrac{1}{2}\,\mathbf{j}\end{aligned}$$

and the position vector of B is as before:

$$\mathbf{r}_B = -2\mathbf{i} + 11\tfrac{1}{2}\,\mathbf{j}$$

So:
$$\begin{aligned}\mathbf{r}_A - \mathbf{r}_B &= (\mathbf{i} + 11\tfrac{1}{2}\,\mathbf{j}) - (-2\mathbf{i} + 11\tfrac{1}{2}\,\mathbf{j}) \\ &= 3\mathbf{i}\end{aligned}$$

The distance between A and B is $|\mathbf{r}_A - \mathbf{r}_B| = 3$ km.

Example 19

A cruiser C is sailing due east at a constant speed of 20 km h^{-1} and a destroyer D is sailing due north at a constant speed of 10 km h^{-1}. At noon C and D are at points with position vectors $(-5\mathbf{i})$ km and $(-20\mathbf{j})$ km, respectively, relative to a fixed origin O. The unit vectors $\mathbf{i}$ and $\mathbf{j}$ are due east and due north respectively.

(a) Show that at time t hours after noon the position vector of C relative to D is given by

$$[(20t - 5)\,\mathbf{i} + (20 - 10t)\,\mathbf{j}]\text{ km}.$$

(b) Show that the distance d km between the vessels, at this time, is given by

$$d^2 = 25\left[20t^2 - 24t + 17\right].$$

(c) Hence show that C and D are closest together at 12.36 p.m. and find the distance between them at this time.

The radar on the cruiser C detects vessels only up to a distance of 15 km.

(d) Determine whether or not the destroyer will be detected by the cruiser's radar.

(a) As the speed of C is constant its velocity is also constant and is $(20\mathbf{i})\,\text{km h}^{-1}$. Similarly the constant velocity of D is $(10\mathbf{j})\,\text{km h}^{-1}$. Proceeding as in example 18, at time t h:

$$\text{position vector } \mathbf{r}_C \text{ km of } C \text{ is } \mathbf{r}_C = -5\mathbf{i} + 20t\mathbf{i} = (20t - 5)\,\mathbf{i}$$

$$\text{position vector } \mathbf{r}_D \text{ km of } D \text{ is } \mathbf{r}_D = -20\mathbf{j} + 10t\mathbf{j} = (10t - 20)\,\mathbf{j}$$

Hence the position vector of C relative to D is:

$$\begin{aligned} \mathbf{r}_C - \mathbf{r}_D &= (20t - 5)\,\mathbf{i} - (10t - 20)\,\mathbf{j} \\ &= (20t - 5)\,\mathbf{i} + (20 - 10t)\,\mathbf{j} \end{aligned}$$

(b) The distance between the two vessels is just $|\mathbf{r}_C - \mathbf{r}_D|$ km and so:

$$\begin{aligned} d^2 &= (20t - 5)^2 + (20 - 10t)^2 \\ &= 25\left[(4t - 1)^2 + (4 - 2t)^2\right] \\ &= 25\left[16t^2 - 8t + 1 + 16 + 4t^2 - 16t\right] \\ &= 25\left[20t^2 - 24t + 17\right] \end{aligned}$$

(c) The expression for d^2 is a quadratic function of t. Quadratic functions are studied in chapter 2 of Book C1, including the method of 'completing the square'. That method can be used to obtain the value of t for which d^2 is a minimum and the corresponding minimum distance.

First, re-write the quadratic function for d^2 so that the coefficient of t^2 is 1. This gives:

$$d^2 = 25 \times 20\left[t^2 - \tfrac{6}{5}t + \tfrac{17}{20}\right]$$

Completing the square gives:

$$\begin{aligned} t^2 - \tfrac{6}{5}t + \tfrac{17}{20} &= \left(t - \tfrac{3}{5}\right)^2 - \left(\tfrac{3}{5}\right)^2 + \tfrac{17}{20} \\ &= \left(t - \tfrac{3}{5}\right)^2 + \tfrac{49}{100} \end{aligned}$$

But $\left(t - \frac{3}{5}\right)^2 \geqslant 0$ and is only zero when $t = \frac{3}{5}$. So the minimum value of the above quadratic occurs when $t = \frac{3}{5}$ and is $\frac{49}{100}$.

The corresponding minimum value of d^2 is

$$25 \times 20 \times \tfrac{49}{100} = 5 \times 49$$

So the two vessels are closest together $\frac{3}{5}$h, that is $\frac{3}{5} \times 60$ minutes $= 36$ minutes, after noon – at 12.36 p.m. The minimum distance is $(5 \times 49)^{\frac{1}{2}} = 7\sqrt{5} = 15.65$ km.

(d) As the minimum distance between the vessels is greater than 15 km – the maximum distance at which the radar can detect vessels – the destroyer will not be detected by the cruiser's radar.

Exercise 2D

In these questions, **i** and **j** are unit vectors in the direction of the positive x- and y-axes respectively. The units of measure are metres and seconds.

1 Points A, B, C, D have coordinates (2,3), (3,−1), (−1,−1) and (0,−4) respectively. Plot these points on coordinate axes Ox and Oy and write down in terms of **i** and **j** the position vector of:

(a) A relative to the origin O
(b) D relative to the origin O
(c) A relative to D
(d) B relative to D
(e) C relative to D.

2 At time $t = 0$ a particle is at the point P, position vector **r**, with velocity vector **v**. Plot the point P on a diagram and draw the velocity vector to show the direction in which the particle is moving in each of the following cases.

(a) $\mathbf{r} = 2\mathbf{i}$, $\mathbf{v} = 2\mathbf{j}$ (b) $\mathbf{r} = 2\mathbf{i}$, $\mathbf{v} = -\mathbf{j}$
(c) $\mathbf{r} = 2\mathbf{j}$, $\mathbf{v} = \mathbf{i} + \mathbf{j}$ (d) $\mathbf{r} = -3\mathbf{i}$, $\mathbf{v} = 3\mathbf{i} + \mathbf{j}$
(e) $\mathbf{r} = \mathbf{i} + \mathbf{j}$, $\mathbf{v} = \mathbf{i} - \mathbf{j}$.

3 For each part of question 2 find the position vector of the particle when $t = 1$, given that the velocity has remained constant between $t = 0$ and $t = 1$.

4 A particle P moving with constant velocity **v** has position vector **a** initially and **b** at time t. Find **v** when:

(a) $\mathbf{a} = 2\mathbf{i}$, $\mathbf{b} = 4\mathbf{i} + 4\mathbf{j}$, $t = 2$
(b) $\mathbf{a} = 2\mathbf{i}$, $\mathbf{b} = 4\mathbf{j}$, $t = 2$
(c) $\mathbf{a} = -2\mathbf{i}$, $\mathbf{b} = 2\mathbf{i} + 8\mathbf{j}$, $t = 4$
(d) $\mathbf{a} = -\mathbf{i} + 3\mathbf{j}$, $\mathbf{b} = 4\mathbf{i} - 2\mathbf{j}$, $t = 5$
(e) $\mathbf{a} = -\mathbf{i} - 4\mathbf{j}$, $\mathbf{b} = 2\mathbf{i} + 2\mathbf{j}$, $t = 3$.

5 A particle has position vector **a** initially and position vector **b** 2 seconds later. Find the velocity vector and the speed of the particle when:

(a) $\mathbf{a} = 3\mathbf{i}$, $\mathbf{b} = 9\mathbf{i}$ (b) $\mathbf{a} = 3\mathbf{i}$, $\mathbf{b} = 9\mathbf{i} + 8\mathbf{j}$
(c) $\mathbf{a} = 3\mathbf{i}$, $\mathbf{b} = -3\mathbf{i} + 8\mathbf{j}$ (d) $\mathbf{a} = -3\mathbf{i}$, $\mathbf{b} = 3\mathbf{i} + 8\mathbf{j}$
(e) $\mathbf{a} = -3\mathbf{i}$, $\mathbf{b} = 5\mathbf{i} + 6\mathbf{j}$.

6 A particle P has velocity $\mathbf{u}$ initially and acceleration $\mathbf{a}$. Find the velocity and the speed after t seconds when:

(a) $\mathbf{u} = 2\mathbf{i}$, $\mathbf{a} = \mathbf{i}$, $t = 3$ (b) $\mathbf{u} = 3\mathbf{i}$, $\mathbf{a} = \mathbf{j}$, $t = 4$

(c) $\mathbf{u} = 5\mathbf{i}$, $\mathbf{a} = 3\mathbf{j}$, $t = 4$ (d) $\mathbf{u} = 5\mathbf{i}$, $\mathbf{a} = -3\mathbf{j}$, $t = 4$

(e) $\mathbf{u} = 8\mathbf{i}$, $\mathbf{a} = 3\mathbf{j}$, $t = 5$.

7 A particle P has velocity $\mathbf{u}$ initially and $\mathbf{v}$ after t seconds. Find its acceleration, assumed constant, when:

(a) $\mathbf{u} = 2\mathbf{i}$, $\mathbf{v} = 4\mathbf{i}$, $t = 4$ (b) $\mathbf{u} = 2\mathbf{i}$, $\mathbf{v} = 4\mathbf{j}$, $t = 2$

(c) $\mathbf{u} = 2\mathbf{i} + 3\mathbf{j}$, $\mathbf{v} = -3\mathbf{i} - 2\mathbf{j}$, $t = 5$

(d) $\mathbf{u} = \mathbf{i} + \mathbf{j}$, $\mathbf{v} = -3\mathbf{i} + 2\mathbf{j}$, $t = 1$

(e) $\mathbf{u} = \mathbf{i} + 2\mathbf{j}$, $\mathbf{v} = 4\mathbf{i} - 3\mathbf{j}$, $t = 1$.

8 A particle has velocity $2\mathbf{i} + 3\mathbf{j}$ initially and velocity $5\mathbf{i} - \mathbf{j}$ 5 seconds later. Find its acceleration, assumed constant, and the magnitude of that acceleration.

9 A particle has position vector $2\mathbf{i} + \mathbf{j}$ initially and is moving with speed $10\,\mathrm{m\,s^{-1}}$ in the direction $3\mathbf{i} - 4\mathbf{j}$. Find its position vector when $t = 3$ and the distance it has travelled in those 3 seconds.

10 A particle has an initial velocity of $\mathbf{i} + 2\mathbf{j}$ and is accelerating in the direction $\mathbf{i} + \mathbf{j}$. If the magnitude of the acceleration is $5\sqrt{2}$, find the velocity vector and the speed of the particle after 2 seconds.

11 (In this question the unit vectors $\mathbf{i}$ and $\mathbf{j}$ are due east and due north respectively.)

At noon a ship S is 600 m due north of an observation point O and a speedboat B is 120 m due north of the same point. The ship S has a constant velocity $(7\mathbf{i} + 8\mathbf{j})\,\mathrm{m\,s^{-1}}$ and the speedboat B has a constant velocity of $(7\mathbf{i} + 24\mathbf{j})\,\mathrm{m\,s^{-1}}$.

(a) Write down the position vectors of S and B at time t seconds after noon.

(b) Show that S and B will collide and find the time when this collision occurs and the position vector of the point of collision.

In order to prevent a collision, 15 s after noon S changes its velocity to $(7\mathbf{i} + 30\mathbf{j})\,\mathrm{m\,s^{-1}}$.

(c) Find the distance between S and B 30 s after noon.

12 At 12 noon the position vectors $\mathbf{r}$ and the velocity vectors $\mathbf{v}$ of two ships A and B are

$\mathbf{r}_A = (2\mathbf{i} + \mathbf{j})\,\text{km}$ $\qquad$ $\mathbf{v}_A = (3\mathbf{i} + \mathbf{j})\,\text{km}\,\text{h}^{-1}$

$\mathbf{r}_B = (-\mathbf{i} - 4\mathbf{j})\,\text{km}$ $\qquad$ $\mathbf{v}_B = (11\mathbf{i} + 3\mathbf{j})\,\text{km}\,\text{h}^{-1}$

(a) Show that at time t h after noon the position vector of B relative to A is given by

$$[(8t - 3)\,\mathbf{i} + (2t - 5)\,\mathbf{j}]\,\text{km}$$

(b) Show that the distance d km between the vessels, at this time, is given by

$$d^2 = 68\left[t^2 - t + \tfrac{1}{2}\right]$$

(c) Hence show that the ships are nearest together at 12.30 p.m. and find the distance between them at this time.

SUMMARY OF KEY POINTS

1 Scalar quantities are completely specified by their magnitude.

2 Vector quantities require both their size (magnitude) and direction to be specified.

3 Vectors are added by using the triangle law of addition:

$\overrightarrow{OB} = \overrightarrow{OA} + \overrightarrow{AB}$

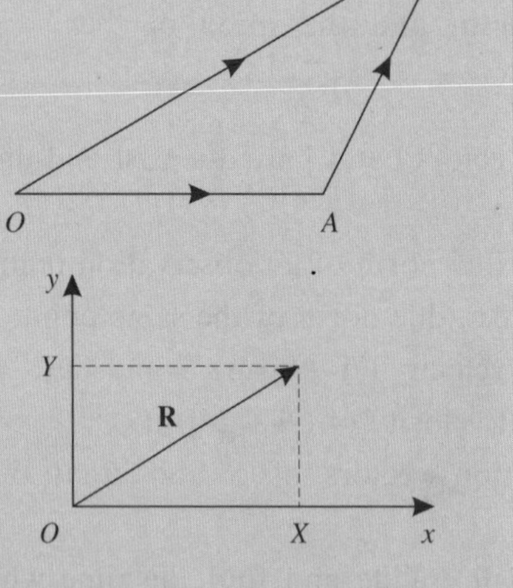

4 Using the $\mathbf{i}$, $\mathbf{j}$ notation:

$\mathbf{R} = X\mathbf{i} + Y\mathbf{j}$

5 If particle A has position vector $\mathbf{r}_A$ and particle B has position vector $\mathbf{r}_B$ then the position vector of B relative to A is $\mathbf{r}_B - \mathbf{r}_A$.

6 If particle A has velocity $\mathbf{v}_A$ and particle B has velocity $\mathbf{v}_B$ then the velocity of B relative to A is $\mathbf{v}_B - \mathbf{v}_A$.

Kinematics of a particle

3

What is kinematics?

Kinematics is the study of the motion of a particle – its speed, acceleration, and the path it follows. Kinematics does not deal with the forces causing the motion but only with the description of the motion itself. (The forces that cause motion are discussed in chapter 5.)

3.1 Motion in a straight line with constant acceleration

Imagine a particle moving in a straight line with a constant or uniform acceleration a for a time interval t. During this time its velocity will change from u, its initial velocity, to v, its final velocity, while it travels a distance s. If you know any three of the quantities a, t, u, v and s you can calculate the remaining two.

The particle's velocity has changed from u to v in time t. Its acceleration is:

$$\text{acceleration} = \frac{\text{change in velocity}}{\text{change in time}}$$

Or:

$$a = \frac{v - u}{t}$$

So:

$$at = v - u$$

■

$$\mathbf{v = u + at} \qquad (1)$$

As the acceleration is constant, the average velocity is the average of the initial and final velocities:

$$\text{average velocity} = \frac{u + v}{2}$$

But the average velocity is also:

$$\text{average velocity} = \frac{\text{displacement}}{\text{time}} = \frac{s}{t}$$

So :

$$\frac{s}{t} = \frac{u + v}{2}$$

■ $$\boldsymbol{s = \left(\frac{u+v}{2}\right)t} \qquad (2)$$

From equation (1), $v = u + at$. Substituting this expression for v into equation (2) gives:

$$s = \left(\frac{u+u+at}{2}\right)t$$

■ $$\boldsymbol{s = ut + \tfrac{1}{2}at^2} \qquad (3)$$

Making t the subject of equation (1) gives:

$$t = \frac{v-u}{a}$$

and substituting this into equation (2) gives:

$$s = \left(\frac{u+v}{2}\right)\left(\frac{v-u}{a}\right)$$

$$2as = uv - u^2 + v^2 - uv$$

$$2as = v^2 - u^2$$

■ $$\boldsymbol{v^2 = u^2 + 2as} \qquad (4)$$

Equations (1) – (4) are very important. You need to remember them, but remember too that they apply to *motion with constant acceleration* only. They are often called the **uniform acceleration equations**.

Example 1

A particle is moving in a straight line from O to A with a constant acceleration of 2 m s^{-2}. Its velocity at A is 30 m s^{-1} and it takes 15 seconds to travel from O to A. Find (a) the particle's velocity at O and (b) the distance OA.

When solving problems of this type it can be useful to start by listing the known quantities. Known quantities are:

$$a = 2\,\text{m}\,\text{s}^{-2}$$

$$v = 30\,\text{m}\,\text{s}^{-1}$$

$$t = 15\,\text{s}$$

(a) To find u, the particle's initial velocity at O, use equation (1), $v = u + at$.

Substituting the known quantities gives:

$$30 = u + 2 \times 15$$

So: $$u = 0$$

The velocity at O is 0 m s^{-1}

(b) To find the distance OA use equation (3), $s = ut + \frac{1}{2}\text{at}^2$.

$$s = 0 + \tfrac{1}{2} \times 2 \times 15^2$$

So: $$s = 225$$

The distance OA is 225 m.

Example 2

A train starts from rest at a station S and moves with constant acceleration. It passes a signal box B 15 seconds later with a speed of 81 km h^{-1}. Modelling the train as a particle find the acceleration of the train in m s^{-2} and the distance in metres between the station and the signal box.

In this example metres, kilometres, hours and seconds are all present. For the uniform acceleration equations to be valid, consistent units must be used. It is usual to work with metres and seconds, so the first step is to change 81 km h^{-1} to m s^{-1}.

As 1 km = 1000 m and 1 hour = 60 × 60 seconds:

$$81\text{ km h}^{-1} = \frac{81 \times 1000}{60 \times 60}\text{ m s}^{-1}$$

$$= 22.5\text{ m s}^{-1}$$

The uniform acceleration equations can now be used.

Known quantities are: $u = 0\text{ m s}^{-1}$

$v = 22.5\text{ m s}^{-1}$

$t = 15\text{ s}$

As: $v = u + at$

$$22.5 = 0 + a \times 15$$

So: $a = \dfrac{22.5}{15} = 1.5$

The acceleration of the train is 1.5 m s^{-2}.

To find s use equation (2), $s = \left(\dfrac{u+v}{2}\right)t$.

Substituting the known quantities gives:

$$s = \left(\frac{0 + 22.5}{2}\right) \times 15$$

So: $s = 168.75$

The distance between the station and the signal box is 169 m.

Retardation (deceleration)

If a particle is slowing down its initial velocity is greater than its final velocity. So its acceleration $\dfrac{v-u}{t}$ is negative. A negative acceleration is called a **retardation** or **deceleration**. To represent a retardation in the uniform acceleration equations you need to use a negative value for the acceleration a, such as -2 m s^{-2}.

Example 3

A boy on a skateboard is travelling up a hill. He experiences a constant retardation of magnitude $2\,\mathrm{m\,s^{-2}}$. Given that his speed at the bottom of the hill was $10\,\mathrm{m\,s^{-1}}$ determine how far he will travel before he comes to rest.

A retardation of $2\,\mathrm{m\,s^{-2}}$ is an acceleration of $-2\,\mathrm{m\,s^{-2}}$.

Known quantities are:

$$a = -2\,\mathrm{m\,s^{-2}}$$
$$u = 10\,\mathrm{m\,s^{-1}}$$
$$v = 0\,\mathrm{m\,s^{-1}}$$

Using: $$v^2 = u^2 + 2as$$

Gives: $$0 = 10^2 + 2 \times (-2)s$$
$$s = \frac{100}{4} = 25$$

The boy travels a distance of 25 m before coming to rest.

Exercise 3A

1 A particle moves in a straight line with uniform acceleration $5\,\mathrm{m\,s^{-2}}$. It starts from rest when $t = 0$. Find its velocity when $t = 3$ s.

2 A particle moves in a straight line. When $t = 0$ its velocity is $3\,\mathrm{m\,s^{-1}}$. When $t = 4$ its velocity is $12\,\mathrm{m\,s^{-1}}$. Find its acceleration, assumed to be constant.

3 A particle moves in a straight line with constant retardation $4\,\mathrm{m\,s^{-2}}$. When $t = 3$ s its velocity is $5\,\mathrm{m\,s^{-1}}$. Find its initial velocity.

4 A particle moves in a straight line with constant acceleration $5\,\mathrm{m\,s^{-2}}$. How long will it take for the particle's velocity to increase from $2\,\mathrm{m\,s^{-1}}$ to $24\,\mathrm{m\,s^{-1}}$?

5 A particle starts from rest and moves in a straight line with constant acceleration $4\,\mathrm{m\,s^{-2}}$ for 3 s. How far does it travel?

6 A particle moves along a line PQ with constant acceleration $3\,\mathrm{m\,s^{-2}}$. If PQ is 2 m and the particle takes 0.5 s to travel from P to Q, what was its velocity at P?

7 A particle travels in a straight line with uniform acceleration $8\,\text{m}\,\text{s}^{-2}$. If its initial velocity is $2\,\text{m}\,\text{s}^{-1}$ how long will it take to travel 6 m?

8 A particle moving in a straight line experiences a constant retardation of $6\,\text{m}\,\text{s}^{-2}$. How long will it take to decrease its speed from $20\,\text{m}\,\text{s}^{-1}$ to $8\,\text{m}\,\text{s}^{-1}$ and how far will it travel while doing so?

9 A particle is moving with uniform acceleration. If it starts from rest and 5 s later has a speed of $18\,\text{m}\,\text{s}^{-1}$ find the distance it has travelled.

10 A particle is moving in a straight line with uniform acceleration. If it travels 120 m while increasing speed from $5\,\text{m}\,\text{s}^{-1}$ to $25\,\text{m}\,\text{s}^{-1}$ find its acceleration.

11 A car is accelerating uniformly while travelling along a straight road. Its speed increases from $8\,\text{m}\,\text{s}^{-1}$ to $22\,\text{m}\,\text{s}^{-1}$ in 10 s. Modelling the car as a particle find the distance travelled during this time and the acceleration of the car.

12 A car is travelling along a straight lane at $20\,\text{m}\,\text{s}^{-1}$ when the driver sees a tractor blocking the lane 30 m ahead. The car's brakes can produce a retardation of $5\,\text{m}\,\text{s}^{-2}$. With what speed does the car hit the tractor?

13 The brakes of a train can produce a retardation of $1.7\,\text{m}\,\text{s}^{-2}$. If the train is travelling at $100\,\text{km}\,\text{h}^{-1}$ and applies its brakes what distance does it travel before stopping? If the driver applies the brakes 15 m too late to stop at a station with what speed is the train travelling when it passes through that station?

14 A particle is moving along a straight line. It passes points A, B and C on the line at $t = 0$, $t = 3$ s and $t = 6$ s respectively. If AC is 60 m and the velocity of the particle at A is $4\,\text{m}\,\text{s}^{-1}$ find the acceleration of the particle (assumed uniform) and the distance AB.

15 A, B and C are three points on a straight road such that $AB = 80$ m and $BC = 60$ m. A car travelling with uniform acceleration passes A, B and C at times $t = 0$, $t = 4$ s and $t = 6$ s respectively. Modelling the car as a particle find its acceleration and its velocity at A.

3.2 Vertical motion under gravity

If you ignore the effect of air resistance on a particle which is moving vertically you can obtain a simplified mathematical model for the particle's motion. This model is, however, a good approximation to the real-life situation. When using this model you are assuming that the particle is moving only in a straight line, that is you are assuming that the object being modelled is not turning or spinning.

The particle is moving with a constant acceleration g due to gravity. This acceleration has a numerical value of approximately $9.8\,\text{m s}^{-2}$. Throughout this book, whenever a numerical value of g is required it will be taken to be $9.8\,\text{m s}^{-2}$.

As the acceleration is constant, the uniform acceleration equations can be used.

When solving a problem, you must choose a direction, either upwards or downwards, to be positive. An arrow indicating the positive direction in your solution will help to remind you.

Example 4

A marble falls off a shelf which is 1.6 m above the floor. Find:

(a) the time it takes to reach the floor
(b) the speed with which it will reach the floor.

In this question the motion is always downwards so choose downwards to be the positive direction. The constant acceleration is in this direction. Since the particle falls off a shelf it will be considered to have an initial speed of $0\,\text{m s}^{-1}$.

Known quantities are: $u = 0\,\text{m s}^{-1}$

↓ $s = 1.6\,\text{m}$

+ve $a = 9.8\,\text{m s}^{-2}$

(a) Use $s = ut + \frac{1}{2}at^2$ to find t.

Substituting the known quantities gives:

$$1.6 = 0 + \tfrac{1}{2} \times 9.8t^2$$

$$t^2 = \frac{1.6 \times 2}{9.8}$$

$$t = 0.5714$$

The time taken to reach the floor is 0.571 s.

(b) Using: $v^2 = u^2 + 2as$

Gives: $v^2 = 0 + 2 \times 9.8 \times 1.6$

$$v = 5.6$$

The marble reaches the floor with a speed of $5.6\,\text{m s}^{-1}$.

- **If a problem involves upward motion, remember the acceleration is always downwards.**

Example 5

A particle P is projected vertically upwards from a point O with a speed of $28\,\text{m s}^{-1}$. Find:

(a) the greatest height h above O reached by P
(b) the total time before P returns to O
(c) the total distance travelled by the particle.

In this question the particle is initially moving upwards and the distance required is measured upwards, so take upwards as the positive direction.

(a) Consider the motion to the highest point:

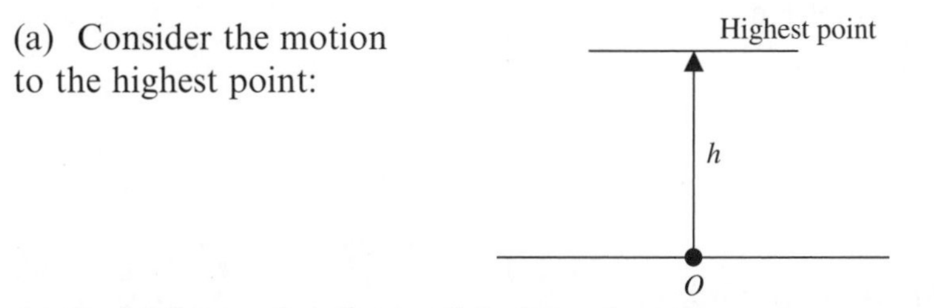

At the highest point the particle is no longer moving upwards but has not yet started to move downwards so its velocity is $0\,\text{m s}^{-1}$.

Known quantities are: $u = 28\,\text{m s}^{-1}$

+ve ↑ $v = 0\,\text{m s}^{-1}$

$a = -9.8\,\text{m s}^{-2}$

$s = h\,\text{m}$

Using: $v^2 = u^2 + 2as$

Gives: $0 = 28^2 + 2 \times (-9.8)h$

$$h = \frac{28^2}{2 \times 9.8} = 40$$

The greatest height above O reached by the particle is 40 m.

(b) When the particle returns to O its displacement from O is zero.

Known quantities are: $s = 0\,\text{m}$

+ve $u = 28\,\text{m s}^{-1}$

↑ $a = -9.8\,\text{m s}^{-2}$

Using: $s = ut + \frac{1}{2}at^2$:

Gives: $0 = 28\,t + \frac{1}{2} \times (-9.8)t^2$

$0 = 28\,t - 4.9\,t^2$

Factorising this quadratic equation as shown in Book P1 chapter 2, gives:

$$0 = (28 - 4.9\,t)t$$

So: $t = 0$ (at start of motion)

Or: $t = \dfrac{28}{4.9} = 5.714$

So the particle returns to its point of projection in 5.71 s.

(c) The total distance travelled by the particle is the distance it travels upwards plus the distance it travels downwards. This is twice the greatest height. The total distance travelled is 40 m.

Example 6

A stone is catapulted vertically upwards with a velocity of $24.5\,\text{m s}^{-1}$. Modelling the stone as a particle moving under gravity alone, find for how long its height exceeds 29.4 m.

Known quantities are: $u = 24.5\,\text{m s}^{-1}$

+ ve $s = 29.4\,\text{m}$

↑ $a = -9.8\,\text{m s}^{-2}$

Using: $s = ut + \frac{1}{2}at^2$

Gives: $29.4 = 24.5\,t - \frac{1}{2} \times 9.8\,t^2$

$$4.9\,t^2 - 24.5\,t + 29.4 = 0$$

÷ 4.9: $t^2 - 5t + 6 = 0$

Factorising, as shown in Book P1 chapter 2, gives:

$$(t - 3)(t - 2) = 0$$

$$t = 2 \text{ or } t = 3$$

The stone is at a height of 29.4 m on its way up at 2 s after projection and again at 3 s, as it comes down.

Its height is greater than 29.4 m for $(3 - 2)\text{s} = 1\,\text{s}$.

Exercise 3B

Whenever a numerical value of g is required take $g = 9.8\,\text{m s}^{-2}$.

1 A ball falls off a table which is 1 m high. Find the speed with which the ball hits the floor.

2 A stone is dropped from the top of a tower which is 20 m high. How long does it take to reach the ground?

3 A book is dropped from the top of a tower. It hits the ground with a speed of $15\,\text{m s}^{-1}$. By modelling the book as a particle find the height of the tower.

4 A ball is thrown vertically upwards with a speed of $10\,\text{m s}^{-1}$. Find (a) the highest point it reaches (b) the total time it is in the air.

5 A stone is catapulted vertically upwards with a velocity of $25\,\text{m s}^{-1}$ from a point 2 m above the ground. Find (a) its velocity when it hits the ground (b) the time it takes to reach the ground.

6 Water from a fountain rises to a height of 6 m. By modelling the drops of water as particles find the speed of the water as it leaves the nozzle.

7 A ball is thrown vertically upwards with a speed of $29\,\text{m s}^{-1}$. It hits the ground 6 seconds later. By modelling the ball as a particle find the height above the ground from which it was thrown.

8 A ball is thrown vertically upwards with a speed of $15\,\text{m s}^{-1}$ from a point 1 m above the ground. Find the speed with which it hits the floor. If it rebounds with a speed which is half the speed with which it hits the floor, find its greatest height after the first bounce.

9 A particle is projected upwards with a speed of $14\,\text{m s}^{-1}$. Find for how long it is above 2 m.

10 A stone is dropped from the top of a tower. One second later another stone is thrown vertically downwards from the same point with a velocity of $14\,\text{m s}^{-1}$. If they hit the ground together find the height of the tower.

vindow which is 30 m above the ground.
particle and hence find the time taken to
at assumptions have you made when
Vhat effect is this likely to have had on

3.3 Speed-time graphs

If a particle is moving with a constant acceleration which causes its speed to change from u to v in time t, a graph showing the particle's change in speed will be a straight line. For speed-time graphs, time is always plotted along the horizontal axis.

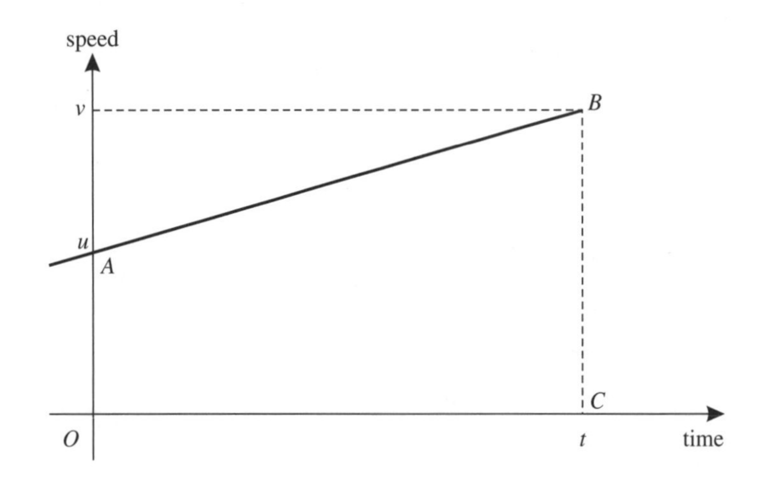

The gradient of the line AB is $\dfrac{v-u}{t}$. This is also the $\dfrac{\text{change in velocity}}{\text{time}}$, that is the acceleration of the particle. Hence the gradient of a speed-time graph gives the acceleration of the particle. $OABC$ is a trapezium with area $\frac{1}{2}(u+v)t$. This is also the distance travelled by the particle as given by the uniform acceleration equations. So the area under the graph is the total distance travelled by the particle for that part of the motion.

- **Gradient of the speed-time graph = acceleration of the particle.**
- **Area under the graph = distance travelled by the particle.**

Example 7

A cyclist leaves home O and rides along a straight road with a constant acceleration. After 10 seconds he has reached point A with a speed of $15\,\text{m}\,\text{s}^{-1}$ and he maintains this speed for a further 20 seconds until he reaches B before retarding uniformly to rest at C. The whole journey takes 45 seconds. Sketch the speed-time graph for the journey and find:

(a) his acceleration for the first part of the journey from O to A
(b) his retardation for the final part of the journey from B to C
(c) the total distance travelled from A to C.

A sketch graph should be drawn on ordinary paper – not on graph paper – but some scales should be put on the axes.

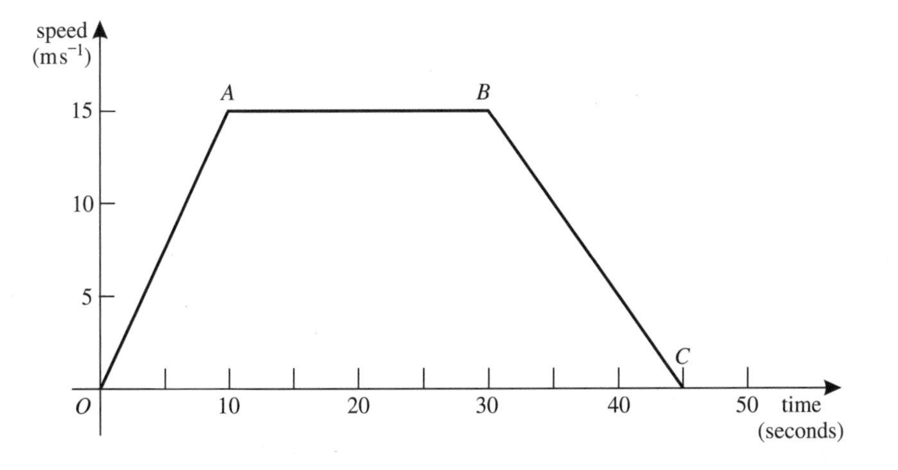

(a) The acceleration for the first part of the journey is given by the gradient of OA.

$$\text{Gradient of } OA = \frac{15}{10} = 1.5$$

The acceleration is $1.5\,\text{m}\,\text{s}^{-2}$.

(b) The acceleration for the final part of the motion is the gradient of BC.

$$\text{Gradient of } BC = \frac{-15}{15} = -1$$

So the retardation is $1\,\text{m}\,\text{s}^{-2}$.

(c) The total distance travelled is given by the area under the graph.

$$\begin{aligned}\text{Area of trapezium } OABC &= \tfrac{1}{2}(20 + 45) \times 15 \\ &= 487.5\end{aligned}$$

The total distance travelled is 488 m.

Example 8

A car travelling along a straight road passes point A when $t = 0$ and maintains a constant speed until $t = 24$ seconds. The driver then applies the brakes and the car retards uniformly to rest at point B. Before the brakes were applied the car had travelled $\frac{4}{5}$ of the total distance AB. Sketch the speed-time graph for the journey and calculate the time taken for the car to travel from A to B.

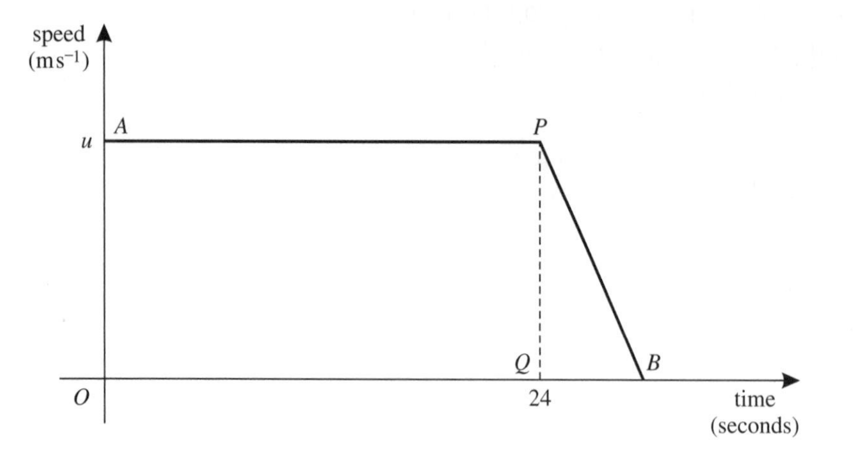

Let the initial speed be $u\,\text{m s}^{-1}$.

Then the distance in metres travelled at constant speed is:

$$\text{Area } OAPQ = (24u)$$

This is $\frac{4}{5}$ of the total distance.

As: $\quad \text{Area } OAPQ = \frac{4}{5} \text{ area } OAPB$

And: $\quad \text{Area } \Delta PBQ = \frac{1}{5} \text{ area } OAPB$

Then:
$$\Delta PBQ = \tfrac{1}{4}\text{area } OAPQ$$
$$= \tfrac{1}{4} \times 24u = 6u$$

Let the car come to rest at time $(24 + T)$ seconds, then QB represents T seconds on the speed-time graph.

Then: $\quad \text{Area } \Delta PBQ = \frac{1}{2}uT$

Hence:
$$\tfrac{1}{2}uT = 6u$$
$$T = 12$$

The total time taken for the car to travel from A to B is $(24 + 12)$ s $= 36$ s.

Example 9

A man is jogging along a straight road at a constant speed of $4\,\text{m s}^{-1}$. He passes a friend with a bicycle who is standing at the side of the road and 20 s later the friend cycles to catch him up. The cyclist accelerates at a uniform rate of $3\,\text{m s}^{-2}$ until he reaches a speed of $12\,\text{m s}^{-1}$. He then maintains a constant speed.

(a) On the same diagram sketch the speed–time graphs for the jogger and the cyclist.
(b) Find the time that elapses before the cyclist reaches the jogger.

(a) For this problem let $t = 0$ at the moment when the jogger passes the cyclist. The speed–time graphs are then as shown below.

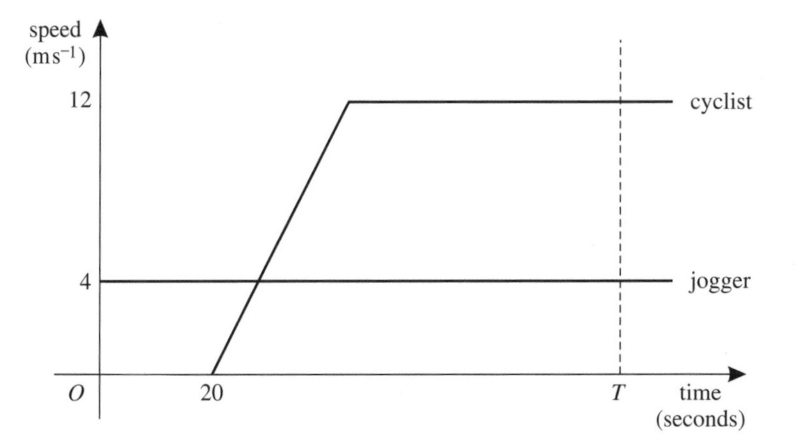

(b) Remember that the jogger and the cyclist will meet when they have travelled the same distance. That is, the areas under their two graphs will be the same.

Let the time when they meet be T s.

Distance travelled by the jogger $= 4T$ m.

The cyclist accelerates at $3\,\text{m s}^{-2}$ to a speed of $12\,\text{m s}^{-1}$. The time for accelerating is thus $\frac{12}{3}$ s $= 4$ s.

The area under the cyclist's graph is a trapezium. So:

$$\text{Distance travelled by the cyclist} \\ = \tfrac{1}{2}[(T - 20) + (T - 24)] \times 12\,\text{m}$$

As the two distances must be the same, we have:

$$4T = \tfrac{1}{2}[(T - 20) + (T - 24)] \times 12$$
$$4T = 6(2T - 44)$$
$$4T = 12T - 6 \times 44$$
$$8T = 6 \times 44$$
$$T = \frac{6 \times 44}{8} = 33$$

The cyclist takes 33 s to catch up with the jogger.

Exercise 3C

1 The speed-time graph below shows the speed of a particle which accelerates uniformly for 2 seconds then moves with constant speed for 6 seconds before retarding to rest. Find (a) the acceleration of the particle (b) the retardation of the particle (c) the distance it travels during the whole motion.

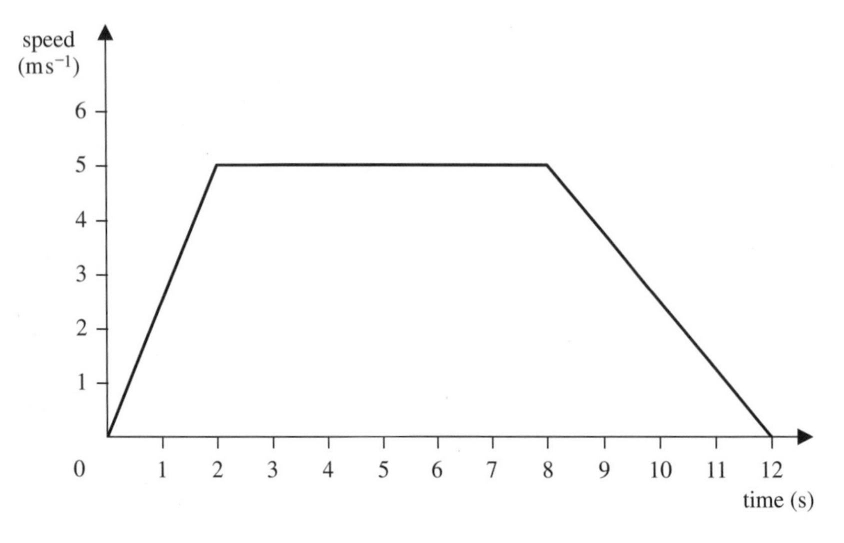

2 A car is moving with a speed of $20\,\mathrm{m\,s^{-1}}$ when the driver sees a red light ahead. He applies the brakes and stops in a distance of 30 m. Sketch a speed-time graph and find (a) the time taken to come to rest (b) the retardation of the car.

3 A car accelerates uniformly from rest with an acceleration of $0.9\,\mathrm{m\,s^{-2}}$ for 12 seconds. It then moves with constant speed for 16 seconds before retarding uniformly to rest. The total distance travelled is 389 m. Show this information on a speed-time graph and find (a) the greatest speed of the car (b) the total time for the journey.

4 Two stations A and B are 495 m apart. A train starts from A and accelerates uniformly for 15 s to a speed of $54\,\mathrm{km\,h^{-1}}$ which it maintains until it is 90 m from B. At this point it slows down uniformly to stop at B. Sketch a speed-time graph to show the motion of the train. Find the magnitude of the acceleration and retardation and the total time the journey takes.

5 A car passes point A on a straight road while travelling at a constant speed of $20\,\text{m s}^{-1}$. Forty seconds later at point B the driver applies the brakes and the car comes to rest at point C. Given that the distance from B to C is $\frac{1}{6}$ of the total distance AC sketch a speed-time graph to show the motion of the car and find the total time taken to travel from A to C.

6 A train starts from rest and moves with uniform acceleration for 5 minutes. It then maintains a constant speed for 20 minutes before being brought to rest by a uniform retardation of magnitude twice that of the acceleration. Sketch a speed-time graph of the motion and find the time it takes the train to stop. If the train travels 4.5 km while accelerating show that the acceleration is $\frac{1}{10}\text{m s}^{-2}$ and find the total distance travelled.

7 A boy is walking at a constant speed of $1.8\,\text{m s}^{-1}$ along a straight road. He passes a telephone booth where his sister is making a telephone call. His sister takes 30 s to complete the call and then sets off in pursuit of the boy. She accelerates uniformly from rest at $3\,\text{m s}^{-2}$ until she is running at a speed of $9\,\text{m s}^{-1}$. She maintains this constant speed until she reaches her brother.

(a) On the same diagram sketch the speed–time graphs for the boy and his sister.

(b) Calculate the time taken by the girl to reach her brother.

8 Two cyclists A and B are travelling in the same direction along a straight road. B is travelling at a speed of $9\,\text{m s}^{-1}$ and A is overtaking B at a speed of $10\,\text{m s}^{-1}$. At the moment when they are level they see a traffic light turn red 108 m ahead. A cycles for T s and then decelerates uniformly; B cycles for 6 s and then decelerates uniformly. A and B stop at the traffic light at the same instant.

(a) Sketch the speed–time graphs of the two cyclists on the same diagram.

(b) Calculate the time during which B was decelerating.

(c) Calculate the value of T.

Exercise 3D Mixed questions

Whenever a numerical value of g is required take $g = 9.8\,\text{m}\,\text{s}^{-2}$.

1 A ball is thrown vertically upwards with speed $10\,\text{m}\,\text{s}^{-1}$ from a point 2 m above horizontal ground.
(a) Calculate the length of time for which the ball is 3 m or more above the ground.
(b) Calculate the speed with which the ball hits the ground.

2 A racing car is moving with constant acceleration along a straight stretch of road. The car passes point A at 0 s, travelling at $18\,\text{m}\,\text{s}^{-1}$, point B at 3 s and point C at 7 s, travelling at $53\,\text{m}\,\text{s}^{-1}$. Calculate
(a) the acceleration of the car,
(b) the speed of the car at B,
(c) the distance from B to C.

3 A train starts from rest at a station and moves along a straight track. Initially the train moves with constant acceleration $3\,\text{m}\,\text{s}^{-2}$ until it is moving with speed $24\,\text{m}\,\text{s}^{-1}$. It maintains this speed for 20 s and then decelerates at $2\,\text{m}\,\text{s}^{-1}$ until it comes to rest at a signal.
(a) Sketch a speed–time graph for the journey.
(b) Calculate the distance from the station to the signal.
(c) Calculate the average speed of the train for the whole journey.

4 A girl throws a ball vertically upwards with speed $8\,\text{m}\,\text{s}^{-1}$ from a window which is 6 m above horizontal ground.
(a) Find the greatest height above the ground reached by the ball.
1.5 s later she drops a second ball from rest out of the same window.
(b) Find the distance below the window of the point where the balls meet.

5 A car starts from rest at traffic lights and moves in a straight line. The car moves with constant acceleration

3 m s^{-2} until it is moving with speed 9 m s^{-1}. It then moves for 5 s with an acceleration of 2 m s^{-2} before moving at a constant speed.

(a) Calculate the constant speed reached by the car.

Given that the car is still moving at this constant speed 12 s after leaving the traffic lights,

(b) calculate the distance travelled in 12 s.

6 A boy drops a large toy car from the top of a cliff which is 90 m above the shore. The car is initially at rest and falls to the shore without hitting anything on the way. By modelling the car as a particle falling freely under gravity, calculate

(a) the time taken for the car to reach the shore,

(b) the speed with which the car hits the shore.

(c) State one factor which has been ignored in modelling the car as described.

7 A train starts from rest at station A and moves with constant acceleration for 60 s until it reaches a speed of 30 m s^{-1}. It travels at this constant speed for T seconds and then decelerates uniformly for 1.2 km, coming to rest at station B which is 14.1 km from station A.

(a) Sketch a speed–time graph for the journey.

(b) Calculate the deceleration of the train.

(c) Calculate the value of T.

(d) Calculate the total time for the journey.

8 A boy drops a stone from rest from the top of a tower which is 25 m high. In an initial model the stone is assumed to be a particle and air resistance is assumed to be negligible.

(a) Calculate the time taken by the stone to reach the ground.

In a refined model air resistance is no longer assumed negligible. Given that in this refined model the stone takes 0.2 s longer to reach the ground,

(b) calculate the revised acceleration of the stone.

SUMMARY OF KEY POINTS

1 **Constant acceleration**

For a particle moving with constant acceleration:

$$v = u + at$$

$$s = \left(\frac{u+v}{2}\right)t$$

$$s = ut + \tfrac{1}{2}at^2$$

$$v^2 = u^2 + 2as$$

2 **Speed-time graphs**

The area under a speed-time graph in a given time interval is the distance travelled during that time.

The gradient of a speed-time graph is the acceleration of the moving particle.

Review exercise 1

Whenever a numerical value of g is required, take $g = 9.8\,\text{m s}^{-2}$.

1 $ABCDEF$ is a regular hexagon with centre O. The vectors $\overrightarrow{OA}$ and $\overrightarrow{OB}$ are denoted by **a** and **b** respectively.

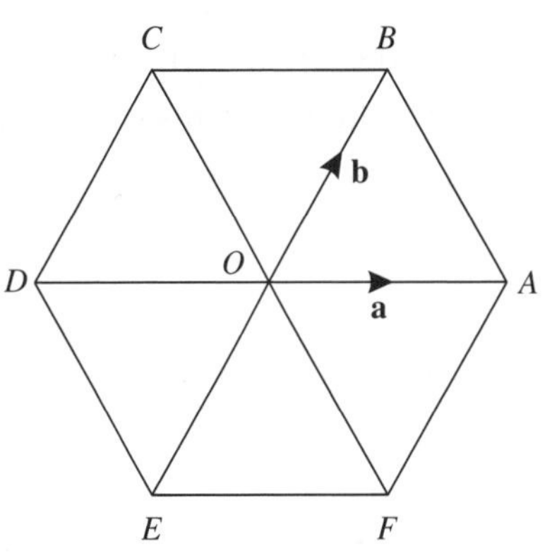

(a) Find in terms of **a** and **b**:

(i) $\overrightarrow{DO}$ (ii) $\overrightarrow{OE}$ (iii) $\overrightarrow{AB}$

(iv) $\overrightarrow{BC}$ (v) $\overrightarrow{FC}$

(b) The vector **b** is equal in magnitude and direction to $\overrightarrow{FA}$ and to $\overrightarrow{DC}$. Using only letters O, A, B, C, D, E, F write down vectors equal to:

(i) $-\mathbf{a}$ (ii) $2\mathbf{a}$ (iii) $\mathbf{a} - \mathbf{b}$

(iv) $\mathbf{b} - \mathbf{a}$ (v) $2(\mathbf{a} - \mathbf{b})$.

2 $ABCDE$ is a pentagon with $\overrightarrow{AB} = \mathbf{p}$, $\overrightarrow{BC} = \mathbf{q}$, $\overrightarrow{AE} = \mathbf{r}$ and $\overrightarrow{ED} = \mathbf{s}$.

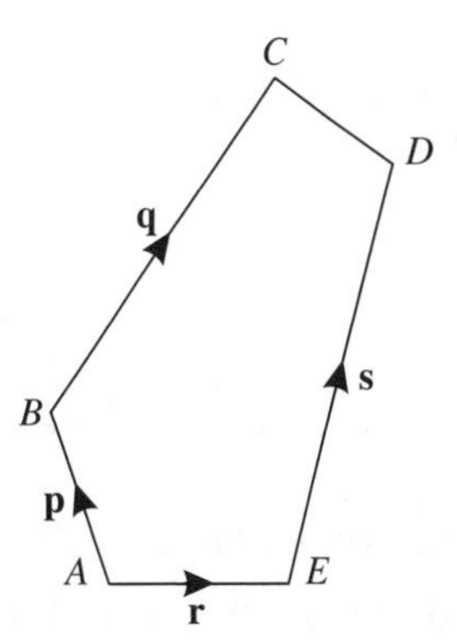

(a) Express in terms of **p**, **q**, **r** and **s**:

(i) $\overrightarrow{BE}$ (ii) $\overrightarrow{CD}$ (iii) $\overrightarrow{EC}$

(iv) $\overrightarrow{AD}$ (v) $\overrightarrow{BD}$.

(b) Given BE is parallel to CD and twice its length, write down an equation in **p**, **q**, **r** and **s**.

(c) Given AD is parallel to BC, write down an equation in **q**, **r**, **s** and a scalar k.

3 OAB is an equilateral triangle. The vectors $\overrightarrow{OA}$ and $\overrightarrow{OB}$ are denoted by **a** and **b** respectively. Copy the figure and mark points whose position vectors relative to O are:

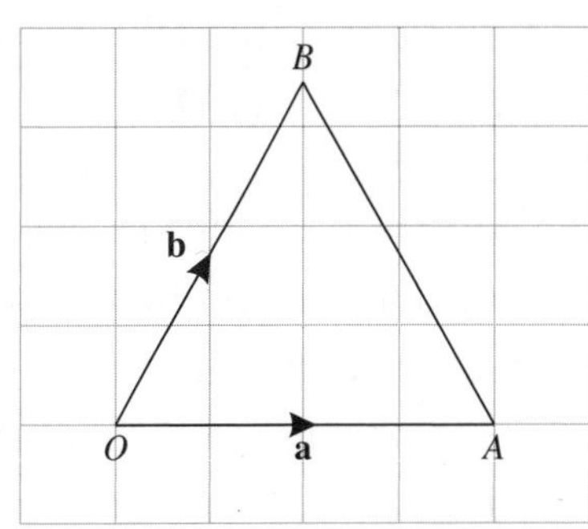

(a) $\mathbf{a} + \mathbf{b}$ (b) $\mathbf{a} - \frac{1}{2}\mathbf{b}$ (c) $\mathbf{a} - 2\mathbf{b}$

(d) $-\mathbf{a} + \mathbf{b}$ (e) $-\mathbf{a} - \frac{1}{2}\mathbf{b}$.

4

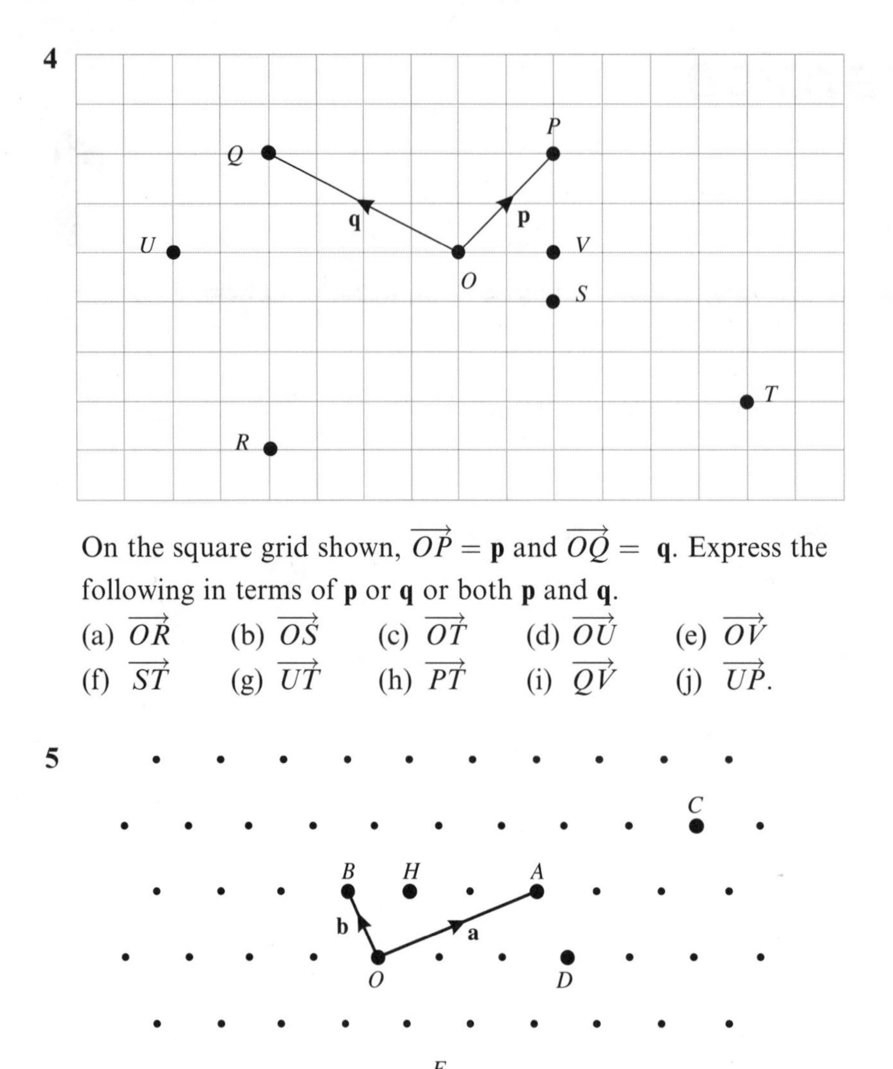

On the square grid shown, $\overrightarrow{OP} = \mathbf{p}$ and $\overrightarrow{OQ} = \mathbf{q}$. Express the following in terms of **p** or **q** or both **p** and **q**.

(a) $\overrightarrow{OR}$ (b) $\overrightarrow{OS}$ (c) $\overrightarrow{OT}$ (d) $\overrightarrow{OU}$ (e) $\overrightarrow{OV}$
(f) $\overrightarrow{ST}$ (g) $\overrightarrow{UT}$ (h) $\overrightarrow{PT}$ (i) $\overrightarrow{QV}$ (j) $\overrightarrow{UP}$.

5

On the triangular grid shown, $\overrightarrow{OA} = \mathbf{a}$ and $\overrightarrow{OB} = \mathbf{b}$. Express the following in terms of **a** or **b** or both **a** and **b**.

(a) $\overrightarrow{OC}$ (b) $\overrightarrow{OD}$ (c) $\overrightarrow{OE}$ (d) $\overrightarrow{OF}$ (e) $\overrightarrow{AE}$
(f) $\overrightarrow{OG}$ (g) $\overrightarrow{OH}$ (h) $\overrightarrow{ED}$ (i) $\overrightarrow{EC}$ (j) $\overrightarrow{FD}$.

6 At time $t = 0$, two ice skaters John (J) and Norma (N) have position vectors $40\mathbf{j}$ metres and $20\mathbf{i}$ metres relative to an origin 0 at the centre of an ice rink, where **i** and **j** are unit vectors perpendicular to each other. John has constant velocity $5\mathbf{i}\,\mathrm{m\,s^{-1}}$ and Norma has constant velocity $(3\mathbf{i} + 4\mathbf{j})\,\mathrm{m\,s^{-1}}$.

(a) Show that the skaters will collide and find the time at which the collision takes place.

(b) On another occasion, John has position vector 40**j** metres. He wishes to skate in a straight line to the point with position vector 30**i** metres. Given that his speed is constant at $5\,\text{m}\,\text{s}^{-1}$, find his velocity.

7 In this question **i** and **j** are horizontal unit vectors and at right angles to each other.
At 12:00 a helicopter A sets out from its base O and flies with speed $120\,\text{km}\,\text{h}^{-1}$ in the direction of the vector $3\mathbf{i} + 4\mathbf{j}$.
(a) Find the velocity vector of A.
At 12:20 that day another helicopter B sets out from O and flies with speed $150\,\text{km}\,\text{h}^{-1}$ in the direction of the vector $24\mathbf{i} + 7\mathbf{j}$.
(b) Find the velocity vector of B.
(c) Find the position vectors of A and B at 13:00.
(d) Calculate the distance of A from B at 13:00.
At 13:30 B makes an emergency landing. A immediately changes direction and flies at $120\,\text{km}\,\text{h}^{-1}$ in a straight line to B.
(e) Find the position vector of B from A at 13:30.
(f) Determine the time when A reaches B.

8 (a) Given that the vector $a\mathbf{i} + 3\mathbf{j}$ is parallel to the vector $2\mathbf{i} + 12\mathbf{j}$ find a.
(b) Given that the vector $3\mathbf{i} + b\mathbf{j}$ is parallel to $12\mathbf{i} - 20\mathbf{j}$ find b.
(c) Given that the vector $7\mathbf{i} + c\mathbf{j}$ has magnitude 25 find the two possible values of c.

9 The position vector **r** of a particle P at time t is given by $\mathbf{r} = t^2\mathbf{i} + (12 - t)\mathbf{j}$. Find the value of t when:
(a) **r** is parallel to the vector **i**
(b) **r** is parallel to the vector $\mathbf{i} + \mathbf{j}$

10 At 11:00 hours the position vector of an aircraft relative to an airport O is $(200\mathbf{i} + 30\mathbf{j})$ km, **i** and **j** being unit vectors due east and due north respectively. The velocity of the aircraft is $(180\mathbf{i} - 120\mathbf{j})\,\text{km}\,\text{h}^{-1}$. Find:
(a) the time when the aircraft is due east of the airport O
(b) how far it then is from O
(c) how far it is from O at 12:00.

11 A train is uniformly retarded from $35\,\text{m s}^{-1}$ to $21\,\text{m s}^{-1}$ over a distance of 350 m. Calculate:

(a) the retardation

(b) the total time taken under this retardation to come to rest from a speed of $35\,\text{m s}^{-1}$. [E]

12 A car uniformly accelerates from rest at $0.7\,\text{m s}^{-2}$ for 6 s. The car then immediately uniformly retards through a distance of 10.5 m and comes to rest. Calculate:

(a) the greatest speed of the car

(b) the time during which the car retards. [E]

13 A particle moves with uniform acceleration $\frac{1}{2}\,\text{m s}^{-2}$ in a horizontal line ABC. The speed of the particle at C is $80\,\text{m s}^{-1}$ and the times taken from A to B and from B to C are 40 s and 30 s respectively. Calculate:

(a) the speed of the particle at A

(b) the distance BC. [E]

14 (a) A car, moving with uniform acceleration along a straight level road, passed points A and B when moving with speed $30\,\text{m s}^{-1}$ and $60\,\text{m s}^{-1}$respectively. Find the speed of the car at the instant it passed C, the mid-point of AB. [E]

(b) State an assumption you have made about the car when forming the mathematical model you used to solve part (a).

15 A car is moving along a straight horizontal road at constant speed $18\,\text{m s}^{-1}$. At the instant when the car passes a lay-by, a motor-cyclist leaves the lay-by, starting from rest, and moves with constant acceleration $2.5\,\text{m s}^{-2}$ in pursuit of the car. Given that the motor-cyclist overtakes the car T seconds after leaving the lay-by, calculate:

(a) the value of T

(b) the speed of the motor-cyclist at the instant of passing the car. [E]

16 A particle P moved in a straight line with constant retardation. At the instants when P passed through the points A, B and C it was moving with speeds $10\,\text{m s}^{-1}$, $7\,\text{m s}^{-1}$ and $3\,\text{m s}^{-1}$ respectively.

Prove that $\frac{AB}{BC} = \frac{51}{40}$. [E]

17 (a) A ball is thrown vertically upwards at $7\,\text{m}\,\text{s}^{-1}$ from a point A which is 8 m vertically above horizontal ground. Given that the ball moves freely under gravity, find:

(i) the greatest height above the ground attained by the ball

(ii) the speed of the ball, in $\text{m}\,\text{s}^{-1}$ to 3 significant figures, at the instant when it strikes the ground. [E]

(b) State two assumptions you have made about the ball and the forces affecting its motion in solving part (a).

18 A stone is thrown vertically upwards with initial speed $28\,\text{m}\,\text{s}^{-1}$. Find the time taken to reach the greatest height that it attains above the point of projection, and find this height. [E]

19 A ball is thrown vertically upwards and takes 3 seconds to reach its highest point. Find the times at which the ball is 39.2 m above its point of projection. [E]

20 A particle starts from rest at O and moves along Ox. During the first 4 seconds of its motion it accelerates uniformly to a speed of $12\,\text{m}\,\text{s}^{-1}$. The particle continues at this speed for 10 seconds. It then decelerates uniformly, coming to rest after a further T seconds. Display this information on a speed-time graph. Given that the total distance travelled is 180 m, calculate T. [E]

21

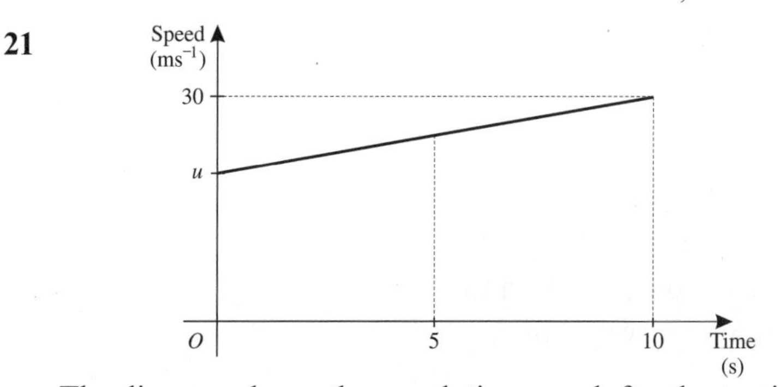

The diagram shows the speed–time graph for the motion of a car which is moving with constant acceleration in a straight line. The car passes a point A with speed $u\,\text{m}\,\text{s}^{-1}$ and, 10 s later, it passes a point B with speed $30\,\text{m}\,\text{s}^{-1}$.

(a) State, in terms of u, the speed of the car at the end of the first 5 seconds.

During the first 5 seconds the car travels 127.5 m.

(b) Calculate the value of u.

(c) Find, in $\text{m}\,\text{s}^{-2}$, the magnitude of the acceleration of the car. [E]

22 A train starts from rest at a station, accelerates uniformly to its maximum speed of $15\,\mathrm{m\,s^{-1}}$, travels at this speed for a time, and then decelerates uniformly to rest at the next station. The distance from station to station is 1260 m, and the time spent travelling at maximum speed is three-quarters of the total journey time.

(a) Illustrate this information on a velocity-time graph.

(b) By using this graph, or otherwise, find the total journey time.

(c) Given also that the magnitude of the deceleration is twice the magnitude of the acceleration find, in $\mathrm{m\,s^{-2}}$, the magnitude of the acceleration. [E]

23 A particle A is constrained to move in a straight line. It starts from rest at X and accelerates at $2\,\mathrm{m\,s^{-2}}$ until it reaches speed V. It then travels with constant speed V for 60 s before decelerating at $1\,\mathrm{m\,s^{-2}}$ to come to rest at Y. The total time for the motion is 105 s. Find the distance XY. [E]

24 A particle starting from rest at time $t = 0$ moves in a straight line and accelerates in the following way:

$$a = 3 \text{ when } 0 \leqslant t \leqslant 20$$
$$a = 1 \text{ when } 20 < t \leqslant 50$$
$$a = -2 \text{ when } 50 < t \leqslant 95$$

where a is the acceleration in $\mathrm{m\,s^{-2}}$ and t is the time in seconds. Find the speed of the particle when $t = 20$, $t = 50$ and $t = 95$. Sketch a time-speed graph for the particle in the interval $0 \leqslant t \leqslant 95$. Find the total distance travelled by the particle in this interval. [E]

25 A particle moves along a straight line. It starts from rest, accelerates at $2\,\mathrm{m\,s^{-2}}$ for 3 seconds, and then decelerates at a constant rate, coming to rest in a further 8 seconds. Sketch a velocity-time graph for this motion. Find the total distance travelled by the particle during these 11 seconds. [E]

26 (a) A cage goes down a vertical mine shaft 650 m deep in 42 s. During the first 16 s it is uniformly accelerated from rest to its maximum speed. For the next 10 s it moves at this maximum speed. It is then uniformly retarded to rest. By drawing a velocity-time graph, or otherwise, find, in m s^{-1}, the maximum speed. [E]

(b) State an assumption you have made about the cage when forming the mathematical model you used to solve (a).

27 The velocities of two particles A and B are $(p\mathbf{i} - 7\mathbf{j})\,\text{m s}^{-1}$ and $(5\mathbf{i} + q\mathbf{j})\,\text{m s}^{-1}$ respectively. The velocity of A relative to B is $(2\mathbf{i} - 3\mathbf{j})\,\text{m s}^{-1}$.
Find the values of p and q. [E]

28 The velocities of two particles A and B are $(13\mathbf{i} - 3\mathbf{j})\,\text{m s}^{-1}$ and $(5\mathbf{i} + 12\mathbf{j})\,\text{m s}^{-1}$ respectively. Find

(a) the speed of B,

(b) the velocity of B relative to A,

(c) the angle between this relative velocity and $\mathbf{i}$, giving your answer to the nearest degree. [E]

29 Two joggers, A and B, are each running with constant velocity on level parkland. At a certain instant, A and B have position vectors $(-60\mathbf{i} + 210\mathbf{j})$ m and $(30\mathbf{i} - 60\mathbf{j})$ m respectively, referred to a fixed origin O. Ninety seconds later, A and B meet at the point with position vector $(210\mathbf{i} + 120\mathbf{j})$ m.

(a) Find, as a vector in terms of $\mathbf{i}$ and $\mathbf{j}$, the velocity of A relative to B.

(b) Verify that the magnitude of the velocity of A relative to B is equal to the speed of A. [E]

30 Two cyclists, C and D, are travelling with constant velocities $(5\mathbf{i} - 2\mathbf{j})\,\text{m s}^{-1}$ and $8\mathbf{j}\,\text{m s}^{-1}$ respectively relative to a fixed origin O.

(a) Find the velocity of C relative to D.

At noon, the position vectors of C and D are $(100\mathbf{i} + 300\mathbf{j})$ m and $(150\mathbf{i} + 100\mathbf{j})$ m respectively, referred to O. At t seconds after noon, the position vector of C relative to D is $\mathbf{s}$ metres.

(b) Show that $\mathbf{s} = (-50 + 5t)\,\mathbf{i} + (200 - 10t)\,\mathbf{j}$.

(c) By considering $|\mathbf{s}|^2$, or otherwise, find the value of t for which C and D are closest together. [E]

31 A car starts from rest at time $t = 0$ seconds and moves with a uniform acceleration of magnitude $2.3\,\text{m}\,\text{s}^{-2}$ along a straight horizontal road. After T seconds, when its speed is $V\,\text{m}\,\text{s}^{-1}$, it immediately stops accelerating and maintains this steady speed until it hits a brick wall when it comes instantly to rest. The car has then travelled a distance of 776.25 m in 30 s.

(a) Sketch a speed–time graph to illustrate this information.

(b) Write down an expression for V in terms of T.

(c) Show that

$$T^2 - 60T + 675 = 0.$$ [E]

32 A car is moving along a straight road with uniform acceleration. The car passes a check-point A with a speed of $12\,\text{m}\,\text{s}^{-1}$ and another check-point C with a speed of $32\,\text{m}\,\text{s}^{-1}$. The distance between A and C is 1100 m.

(a) Find the time, in seconds, taken by the car to move from A to C.

Given that B is the mid-point of AC,

(b) find, in $\text{m}\,\text{s}^{-1}$ to 1 decimal place, the speed with which the car passes B. [E]

33 A train stops at two stations 7.5 km apart. Between the stations it takes 75 s to accelerate uniformly to a speed of $24\,\text{m}\,\text{s}^{-1}$, then travels at this speed for a time T seconds before decelerating uniformly for the final 0.6 km.

(a) Sketch a speed–time graph for this journey.

Hence, or otherwise, find

(b) the deceleration, in $\text{m}\,\text{s}^{-2}$, of the train during the final 0.6 km,

(c) the value of T,

(d) the total time for the journey.

34 A particle moves with constant acceleration along the straight line OLM and passes through the points O, L and M at times 0 s, 4 s and 10 s respectively. Given that $OL = 14$ m and $OM = 50$ m, find

(a) the acceleration of the particle,

(b) the speed of the particle at M.

Statics of a particle

4

'Static' means 'stationary' or 'at rest', and this chapter looks at the forces acting on a particle that is not moving. Later in this chapter the various types of forces that may act on a particle will be considered. At this point you only need to remember from chapter 2 that you can describe a force acting on a particle by stating its magnitude and its direction.

Forces are measured in **newtons**. The precise definition of a newton is given in chapter 5 (p. 92). A force of magnitude of 8 newtons is usually written 8 N.

4.1 Resultant forces

Suppose two forces $\mathbf{F}_1$ and $\mathbf{F}_2$ act on a particle as shown. As forces are vectors they may be added by the triangle law (see p. 15) You can redraw the diagram as:

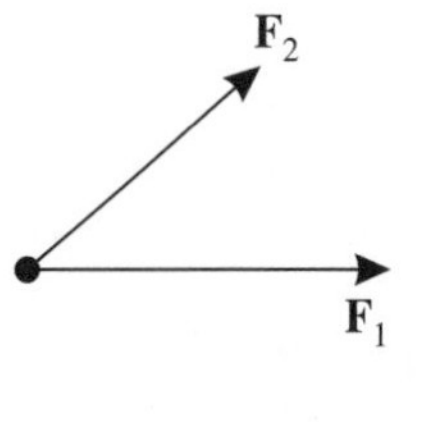

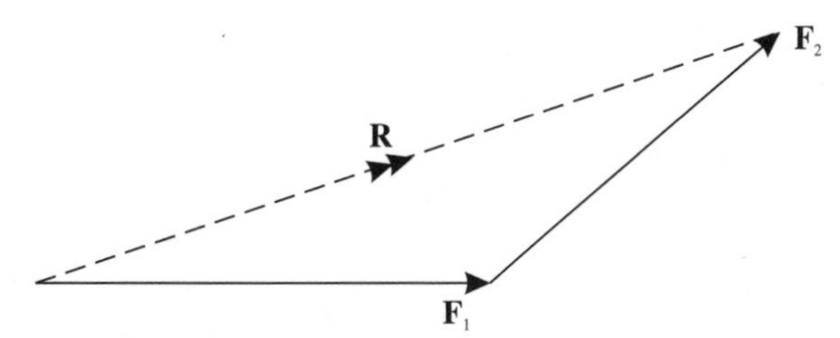

Then $\mathbf{R} = \mathbf{F}_1 + \mathbf{F}_2$ is the **resultant** of the two forces. Use a double arrow to indicate that $\mathbf{R}$ is a resultant force rather than a third force acting on the particle.

Another way of adding forces is to use the **parallelogram rule**. (See p. 15).

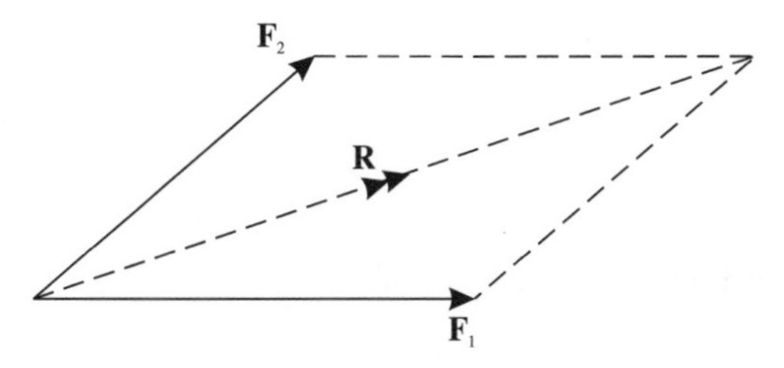

The resultant **R** is then given by the diagonal of the parallelogram. The advantage of this method is that it shows the two forces and their resultant acting at the same point. However, it is usually more convenient in calculations to use the triangle law.

Example 1

Two forces of magnitude 5 N and 6 N act on a particle. They act at right angles. Find the magnitude and direction of their resultant.

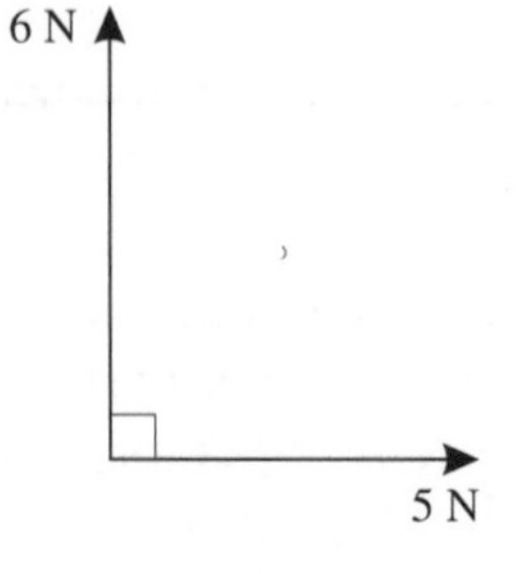

Let R N be the magnitude of the resultant and θ the angle it makes with the 5 N force. The vector triangle is:

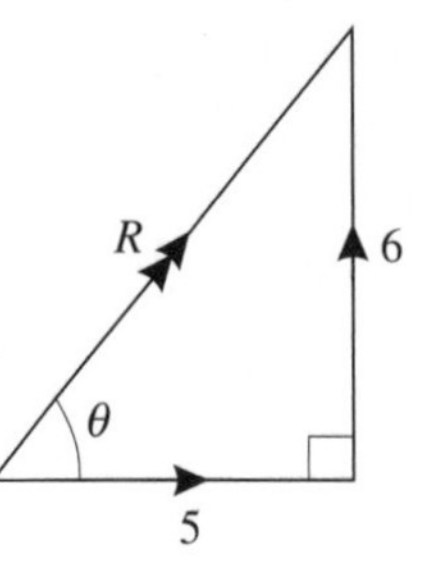

By Pythagoras' Theorem:

$$R^2 = 6^2 + 5^2 = 61$$
$$R = 7.81$$

Also:

$$\tan\theta = \frac{6}{5} = 1.2$$
$$\theta = 50.2^\circ$$

The resultant is a force of magnitude 7.81 N at an angle of 50.2° to the 5 N force.

Example 2

Two forces acting on a particle have magnitudes 8 N and 3 N. The angle between their directions is 60°. Find the resultant force acting on the particle.

The force diagram is:

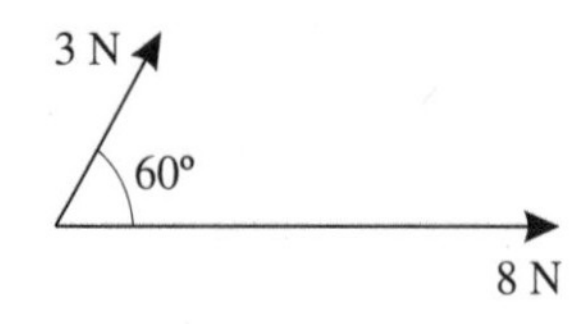

The vector triangle for this situation is:

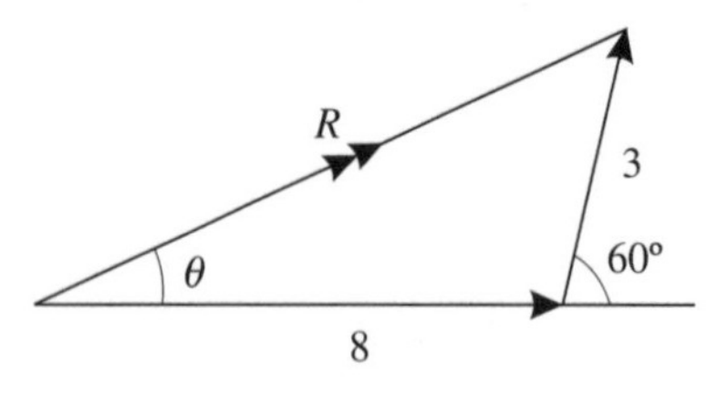

Let the resultant be of magnitude R N and make an angle of θ with the 8 N force.

By the cosine rule: $R^2 = 3^2 + 8^2 - 2 \times 8 \times 3 \times \cos 120°$
$= 9 + 64 - 48(-\frac{1}{2})$
$= 97$
$R = \sqrt{97} = 9.85$

By the sine rule: $\dfrac{R}{\sin 120°} = \dfrac{3}{\sin \theta}$

Giving: $\sin \theta = \dfrac{3 \sin 120°}{R}$
$= \dfrac{3 \sin 120°}{\sqrt{97}}$

So: $\theta = 15.3°$

The resultant force is of magnitude 9.85 N at an angle of 15.3° to the 8 N force.

(Notice that to achieve maximum accuracy in θ you should use $\sqrt{97}$ for R and not the corrected value of 9.85.)

Finding the resultant of more than two forces

The resultant $\mathbf{R}_1$ of any two forces **P** and **Q** acting on a particle can be found by constructing a vector triangle.

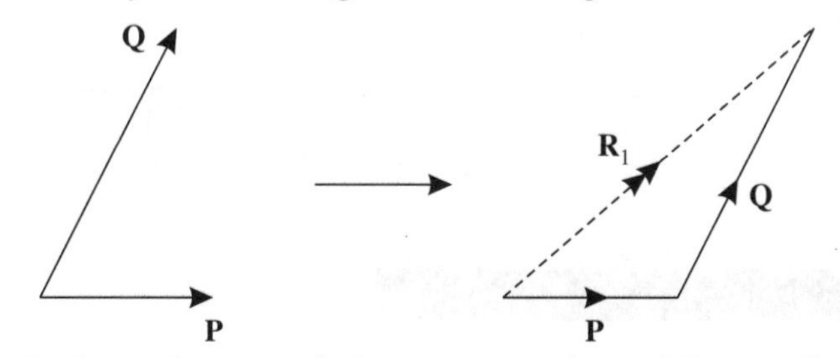

This method can be extended to any number of forces. Suppose a third force **S** also acts on the particle. The resultant $\mathbf{R}_2$ of the three forces is then the resultant of $\mathbf{R}_1$ and **S**. You can find $\mathbf{R}_2$ by drawing a **polygon of forces**: the resultant $\mathbf{R}_2$ is represented by the line required to complete the polygon.

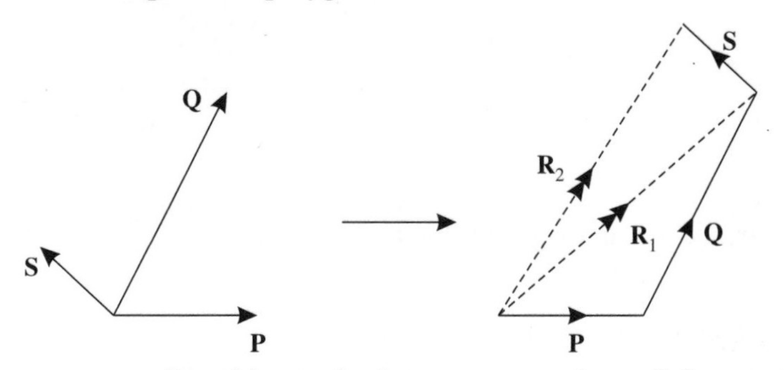

You can generalise this method to any number of forces. The magnitude and direction of the resultant can be calculated using trigonometry.

However, it is easier to find the resultant of a set of forces if the forces are expressed in terms of cartesian components or resolutes (see section 2.3) In the following example the forces are given in the **i**, **j** notation.

Example 3

Three forces $(2\mathbf{i} - \mathbf{j})$ N, $3\mathbf{i}$ N and $(-\mathbf{i} + 4\mathbf{j})$ N, where **i** and **j** are unit vectors due east and due north respectively, act on a particle. Find the magnitude and direction of the resulting force.

The forces are given in the **i**, **j** notation, so you can add the vectors by the method given in section 2.3 (p. 20) to obtain the resultant **R** N:

$$\begin{aligned}\mathbf{R} &= (2\mathbf{i} - \mathbf{j}) + 3\mathbf{i} + (-\mathbf{i} + 4\mathbf{j}) \\ &= 4\mathbf{i} + 3\mathbf{j}\end{aligned}$$

The magnitude of the resultant is:

$$|\mathbf{R}| = \sqrt{(4^2 + 3^2)} = \sqrt{25} = 5$$

The diagram shows that:

$$\begin{aligned}\tan\theta &= \tfrac{3}{4} \\ \theta &= 36.9^\circ\end{aligned}$$

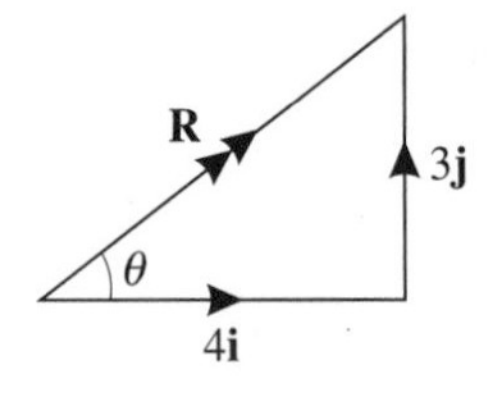

The resultant is therefore of magnitude 5 N and acts at an angle of 36.9° to unit vector **i**.

Exercise 4A

1 Forces **P** and **Q** act on a particle at the origin O along the coordinates axes Ox, Oy respectively. Calculate the magnitude of the resultant force and the angle it makes with the x-axis when:

(a) $P = 4$ N, $Q = 3$ N
(b) $P = 5$ N, $Q = 8$ N
(c) $P = 8$ N, $Q = 5$ N
(d) $P = 24$ N, $Q = 15$ N
(e) $P = 9$ N, $Q = 40$ N.

2 Two forces **P** and **Q** act on a particle at the origin O. Force **P** acts along Ox, force **Q** makes an angle θ with Ox.
Calculate the magnitude of the resultant force and the angle it makes with Ox when:

(a) $P = 4$ N, $Q = 3$ N, $\theta = 60^\circ$
(b) $P = 3$ N, $Q = 4$ N, $\theta = 60^\circ$

(c) $P = 5\,\text{N}$, $Q = 5\,\text{N}$, $\theta = 60°$
(d) $P = 10\,\text{N}$, $Q = 12\,\text{N}$, $\theta = 120°$
(e) $P = 8\,\text{N}$, $Q = 5\,\text{N}$, $\theta = 120°$.

3 Forces **P**, **Q** and **R** act on a particle at O in the plane of the coordinate axes Ox, Oy. Force **P** acts along Ox, **Q** acts along Oy, and **R** acts at an angle θ with Ox, in the first quadrant. Calculate the magnitude of the resultant force and the angle it makes with Ox when:
(a) $P = 3\,\text{N}$, $Q = 4\,\text{N}$, $R = 5\,\text{N}$, $\theta = 60°$
(b) $P = 4\,\text{N}$, $Q = 3\,\text{N}$, $R = 5\,\text{N}$, $\theta = 60°$
(c) $P = 5\,\text{N}$, $Q = 6\,\text{N}$, $R = 7\,\text{N}$, $\theta = 45°$
(d) $P = 6\,\text{N}$, $Q = 6\,\text{N}$, $R = 6\,\text{N}$, $\theta = 60°$
(e) $P = 6\,\text{N}$, $Q = 6\,\text{N}$, $R = 6\,\text{N}$, $\theta = 45°$.

4 Forces **P**, **Q** and **R** act on a particle at the origin O. Find the resultant of the three forces in terms of **i** and **j**, the unit vectors along Ox and Oy respectively, when:
(a) $\mathbf{P} = 2\mathbf{i} + 5\mathbf{j}$, $\mathbf{Q} = 3\mathbf{i} + 4\mathbf{j}$, $\mathbf{R} = 5\mathbf{i} + 2\mathbf{j}$
(b) $\mathbf{P} = 2\mathbf{i} + 5\mathbf{j}$, $\mathbf{Q} = 3\mathbf{i}$, $\mathbf{R} = 5\mathbf{j}$
(c) $\mathbf{P} = 2\mathbf{i} + 5\mathbf{j}$, $\mathbf{Q} = 2\mathbf{i} - 5\mathbf{j}$, $\mathbf{R} = 3\mathbf{i}$
(d) $\mathbf{P} = 2\mathbf{i} + 5\mathbf{j}$, $\mathbf{Q} = \mathbf{i} - 7\mathbf{j}$, $\mathbf{R} = -3\mathbf{i} - 4\mathbf{j}$
(e) $\mathbf{P} = 2\mathbf{i} + 5\mathbf{j}$, $\mathbf{Q} = -3\mathbf{i} - 4\mathbf{j}$, $\mathbf{R} = \mathbf{i} - \mathbf{j}$.

5 For each part of question 4 find the magnitude of the resultant force and the angle it makes with Ox.

6 Forces **P**, **Q**, **R** and **S** act on a particle at O in the plane of the coordinate axes Ox, Oy, making angles p, q, r, s respectively with Ox, each angle being measured in the anticlockwise sense. By drawing a polygon of forces find the magnitude of their resultant and the angle it makes with Ox when:
(a) $P = 2\,\text{N}$, $Q = 3\,\text{N}$, $R = 4\,\text{N}$, $S = 5\,\text{N}$, $p = 0°$, $q = 40°$, $r = 100°$, $s = 150°$
(b) $P = 3\,\text{N}$, $Q = 3\,\text{N}$, $R = 5\,\text{N}$, $S = 5\,\text{N}$, $p = 0°$, $q = 40°$, $r = 130°$, $s = 220°$
(c) $P = 1\,\text{N}$, $Q = 2\,\text{N}$, $R = 3\,\text{N}$, $S = 4\,\text{N}$, $p = 10°$, $q = 70°$, $r = 100°$, $s = 300°$.

eful to resolve a
a very important
th are forces. In
esolved parts of

the coordinate

useful to find
ther times it is
lel to and at right angles
d choose the directions for the
duce the simplest possible equations. This is
d by the direction of forces such as the weight of the
particle and the frictional force.

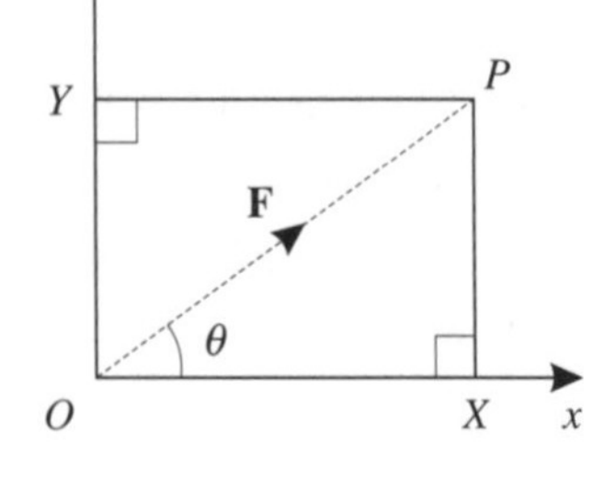

Suppose the force **F** acts at an angle θ to the x-axis.
By the triangle law:

$$\overrightarrow{OP} = \overrightarrow{OX} + \overrightarrow{XP}$$
$$= \overrightarrow{OX} + \overrightarrow{OY}$$

Then the components or resolutes of **F** along the x- and y-axes are $\overrightarrow{OX}$ and $\overrightarrow{OY}$.

By trigonometry:

$$\frac{OX}{OP} = \cos\theta$$

So: $$OX = OP\cos\theta = F\cos\theta$$

And: $$\frac{OY}{OP} = \sin\theta$$

So: $$OY = OP\sin\theta = F\sin\theta$$

So the components of **F** along the x- and y-axes are:

$$F\cos\theta \qquad \text{and} \qquad F\sin\theta$$

and: $$\mathbf{F} = F\cos\theta\,\mathbf{i} + F\sin\theta\,\mathbf{j}$$

As $\sin\theta = \cos(90° - \theta)$ you can write this as:

$$\mathbf{F} = F\cos\theta\,\mathbf{i} + F\cos(90° - \theta)\mathbf{j}$$

F makes angles θ and $(90° - \theta)$ with the x- and y-axes respectively. This leads to a general rule for obtaining components or resolutes of a force:

- **The component of a force in any direction is the product of the magnitude of the force and the cosine of the angle between the force and the required direction.**

Notice in particular that the component of a force in a direction perpendicular to that force is zero (as $\theta = 0°$ in that case and $\cos(90-0)° = \cos 90° = 0$).

Example 4

A force of magnitude 5 N acts at an angle of 62° to the horizontal. Find the magnitudes of the horizontal and vertical components of this force.

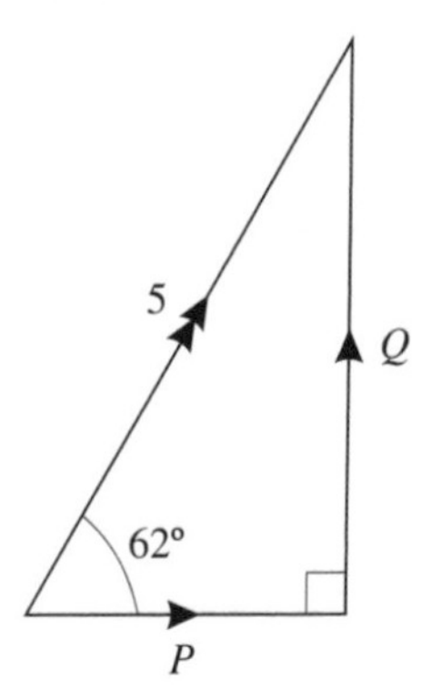

Let P and Q be the horizontal and vertical components of the force.

$$P = 5 \cos 62° = 2.35$$
$$Q = 5 \cos (90-62)°$$

Or more usually:

$$Q = 5 \sin 62° = 4.14$$

So the magnitudes of the horizontal and vertical components are 2.35 N and 4.14 N.

Example 5

A particle lies on the face of a plane inclined at an angle θ to the horizontal. It is acted on by a force **R** perpendicular to the plane and a force **W** vertically downwards. Find the components of the resultant force on the particle:

(a) horizontally and vertically

(b) down the plane and perpendicular to the plane.

The force diagram for this situation is:

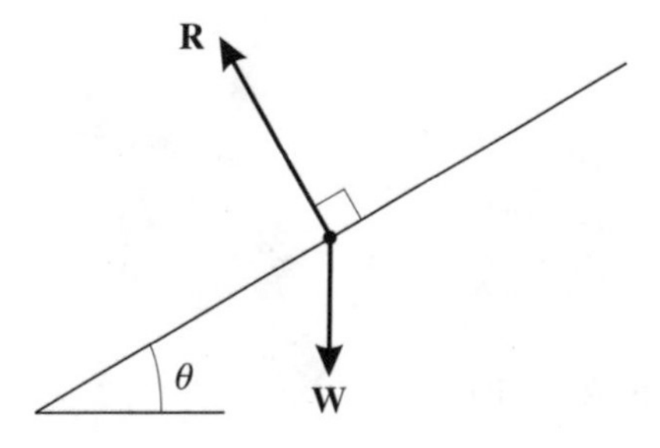

Before attempting to obtain components it is necessary to work out some important angles. The diagram below shows the essentials.

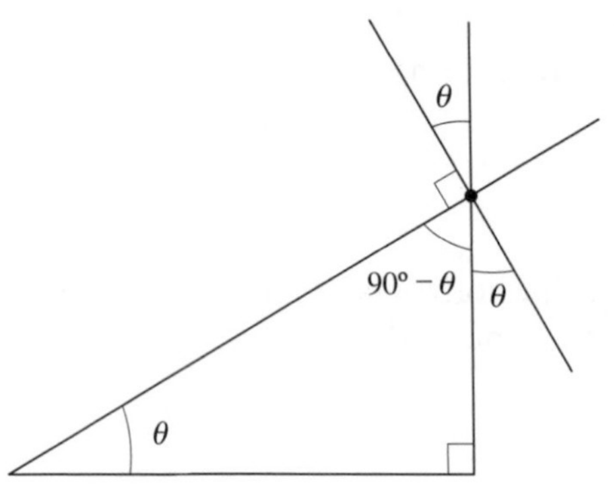

(a) Resolve the force **R** into horizontal and vertical components:

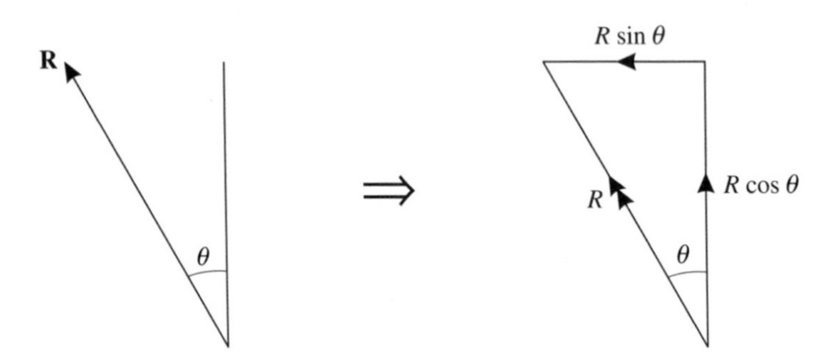

The force **W** is vertical, so it has *no* horizontal component. So:

horizontal component of the resultant is $R \sin \theta \leftarrow$
vertical component of the resultant is $R \cos \theta - W \uparrow$

(b) Clearly **R** will have no component down the plane as **R** is perpendicular to this direction.
Resolving **W** in these directions gives:

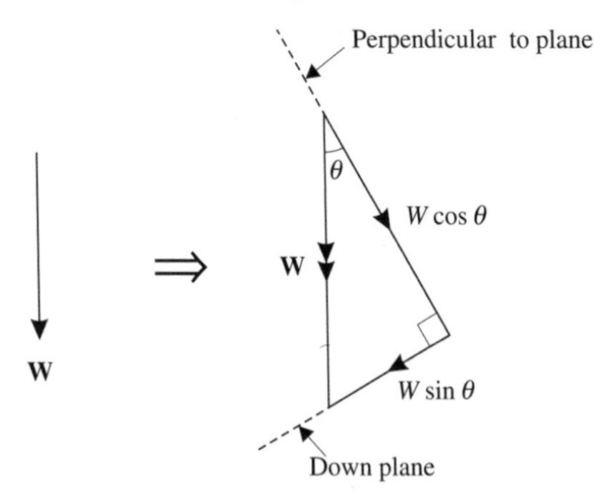

So:

component of resultant down plane is $W \sin \theta \swarrow$
component of resultant perpendicular to plane is $R - W \cos \theta \nwarrow$

The arrows are added to make it quite clear in which direction each component force acts.

When you combine the components to find the resultant vector, you must be careful to show the components in the correct direction in your force diagram.

4.3 Finding the resultant of several forces by resolving them into components

Suppose several forces act on a particle. One way to work out the resultant of the forces is to find the components of each force in two perpendicular directions. You can then find the algebraic sum of these components to give the magnitude and direction of the resultant.

Example 6

Here are four forces acting on a particle. Find the magnitude and direction of the resultant force.

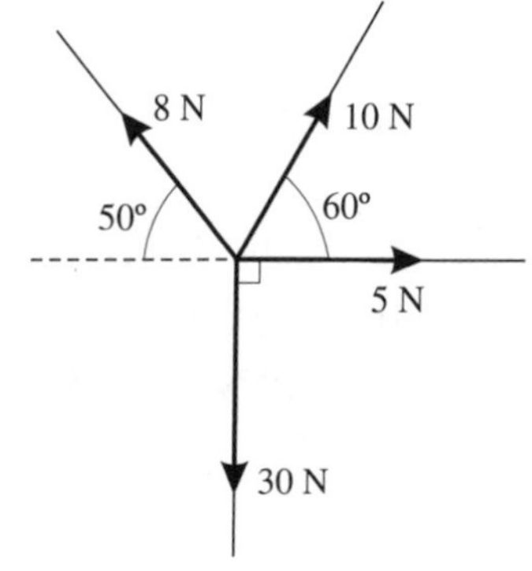

The question does not tell you which directions you should choose when you resolve the forces into components. Take $\rightarrow$ along the 5 N force and $\downarrow$ along the 30 N force. These directions are perpendicular. The 30 N force will not contribute anything to the horizontal component, and the 5 N force will not contribute to the vertical component.

Call the components of the resultant $X \rightarrow$ and $Y \downarrow$.

Resolving horizontally $\rightarrow$ gives: $X = 5 + 10\cos 60^\circ - 8\cos 50^\circ$

$$= 4.86$$

Resolving vertically $\downarrow$ gives: $Y = 30 - 10\sin 60^\circ - 8\sin 50^\circ$

$$= 15.2$$

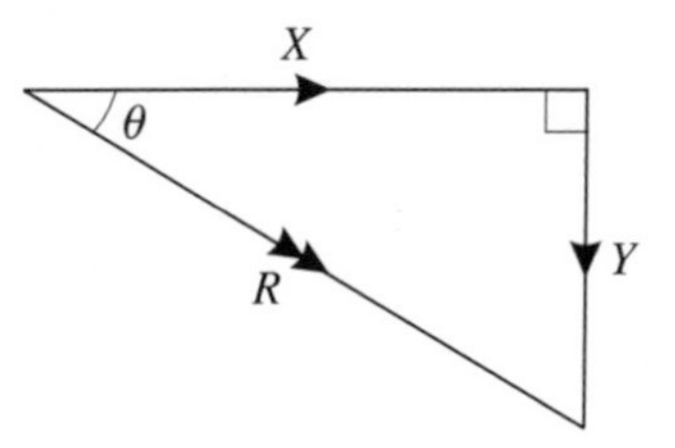

So:

$$R = \sqrt{(X^2 + Y^2)}$$
$$= \sqrt{(4.86^2 + 15.2^2)}$$
$$= 16.0$$

and:

$$\tan\theta = \frac{Y}{X} = \frac{15.2}{4.86}$$
$$\theta = 72.3^\circ$$

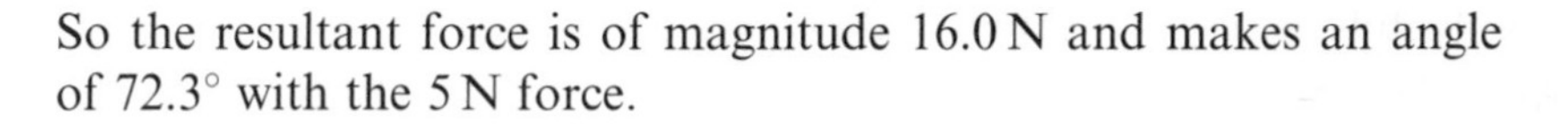

So the resultant force is of magnitude 16.0 N and makes an angle of 72.3° with the 5 N force.

Exercise 4B

1 A force **F** acts on a particle at the origin O in the plane of the coordinate axes Ox, Oy, at an angle θ to Ox, θ being measured in the anticlockwise sense. Find the components of **F** along Ox and Oy when:

(a) $F = 4\,\text{N}$, $\theta = 30°$
(b) $F = 6\,\text{N}$, $\theta = 60°$
(c) $F = 6\,\text{N}$, $\theta = 120°$
(d) $F = 8\,\text{N}$, $\theta = 150°$
(e) $F = 10\,\text{N}$, $\theta = 240°$.

2 A force **F** acts on a particle at an angle of θ to the horizontal. Find the horizontal and vertical components of **F** when:

(a) $F = 2\,\text{N}$, $\theta = 20°$
(b) $F = 10\,\text{N}$, $\theta = 30°$
(c) $F = 20\,\text{N}$, $\theta = 50°$

3 A force **W** acts vertically downwards on a particle at rest on a plane inclined at an angle θ to the horizontal. Find the components of **W** along and perpendicular to the plane when:

(a) $W = 2\,\text{N}$, $\theta = 30°$
(b) $W = 4\,\text{N}$, $\theta = 50°$
(c) $W = 10\,\text{N}$, $\theta = 75°$.

4

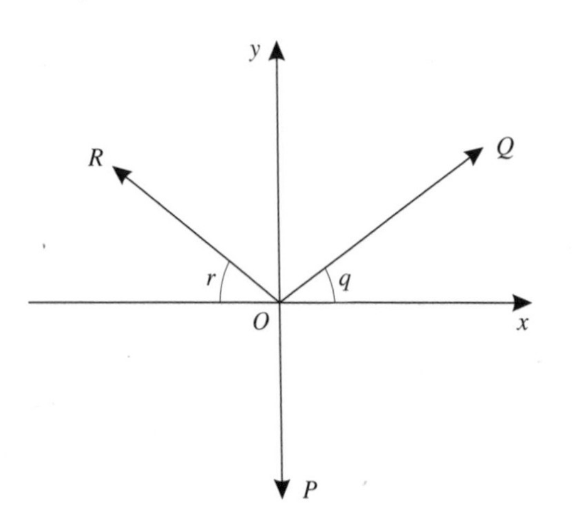

Forces **P**, **Q** and **R** act on a particle at rest at O, as shown in the diagram. By first finding the components of each force along Ox and Oy, find the magnitude of the resultant force and the angle it makes with Ox when:

(a) $P = 5\,\text{N}$, $Q = 4\,\text{N}$, $R = 3\,\text{N}$, $q = 20°$, $r = 40°$
(b) $P = 5\,\text{N}$, $Q = 6\,\text{N}$, $R = 4\,\text{N}$, $q = 20°$, $r = 40°$
(c) $P = 10\,\text{N}$, $Q = 4\,\text{N}$, $R = 5\,\text{N}$, $q = 30°$, $r = 60°$.

5

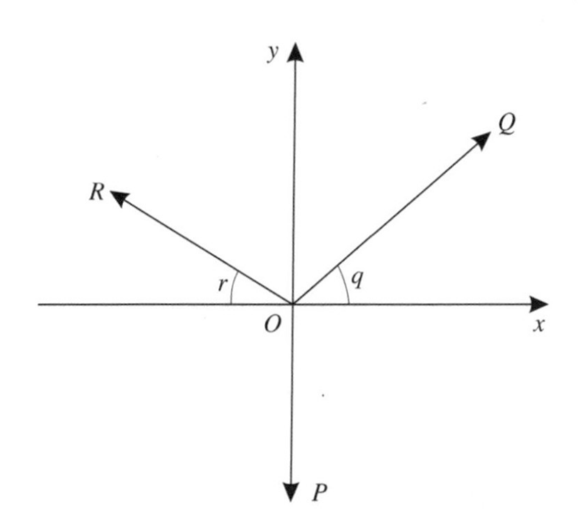

Forces P, Q and R act on a particle at rest at the origin O, as shown in the diagram. By first finding the components of each force along the line of action of Q, find the magnitude of the resultant force and the angle its line of action makes with the line of action of Q when:
(a) $P = 2\,\text{N}$, $Q = 3\,\text{N}$, $R = 4\,\text{N}$, $q = 20°$, $r = 70°$
(b) $P = 5\,\text{N}$, $Q = 6\,\text{N}$, $R = 7\,\text{N}$, $q = 30°$, $r = 60°$
(c) $P = 10\,\text{N}$, $Q = 20\,\text{N}$, $R = 30\,\text{N}$, $q = 60°$, $r = 30°$.

4.4 Equilibrium of coplanar forces

If all the forces acting on a particle are in the same plane they are called **coplanar forces**. This book deals only with coplanar forces. In the real world, of course, forces can act in all directions on a particle. If all the forces acting on a particle cancel each other out, so that nothing happens at all, the forces are said to be in **equilibrium**.

- **A system of forces acting on a particle is said to be in equilibrium if their resultant is the zero vector.**

When forces in equilibrium are resolved into components in two fixed directions, the algebraic sum of the components in each direction will also be zero. This is the basic fact you need to know to deal with particles in equilibrium.

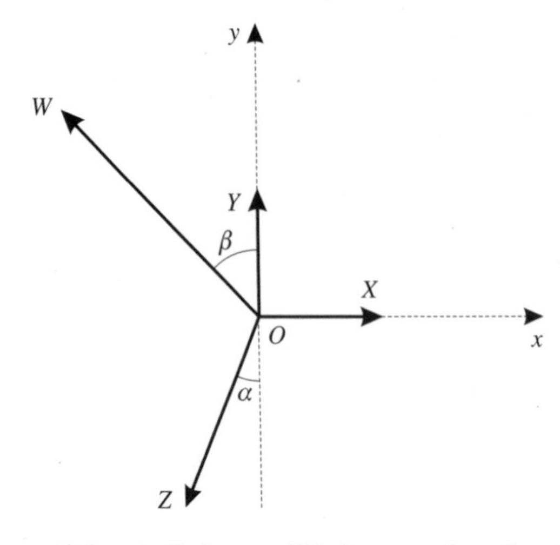

Consider a particle at O in equilibrium under the action of the forces **X**, **Y**, **W** and **Z** as shown above.

Resolving horizontally → gives: $X - W \sin \beta - Z \sin \alpha$

Resolving vertically ↑ gives: $Y + W \cos \beta - Z \cos \alpha$

As the particle is in equilibrium, each of these expressions must be equal to zero.

So: $X - W \sin \beta - Z \sin \alpha = 0$

and: $Y + W \cos \beta - Z \cos \alpha = 0$

There are two equations, so you can use them to find two unknown quantities. Whichever two directions you choose, you will always get two equations like these.

Example 7

A particle is in equilibrium under the action of the forces shown. Find the magnitude of the forces **P** and **S**.

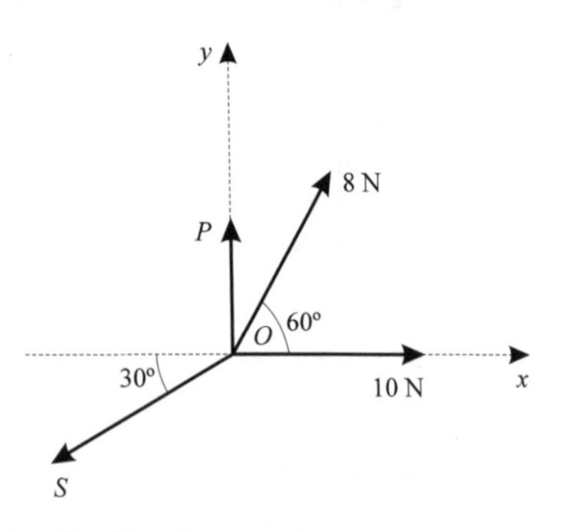

Resolving in the Ox direction → gives:
$10 + 8 \cos 60° - S \cos 30° = 0$

As $\cos 60^\circ = \frac{1}{2}$ and $\cos 30^\circ = \frac{\sqrt{3}}{2}$ (see Book P1, section 3.4):

$$10 + 8 \times \tfrac{1}{2} - S \times \frac{\sqrt{3}}{2} = 0$$

$$14 - S \times \frac{\sqrt{3}}{2} = 0$$

$$S = \frac{28}{\sqrt{3}} = \frac{28\sqrt{3}}{3}$$

Resolving in Oy direction $\uparrow$ gives: $P + 8 \sin 60^\circ - S \sin 30^\circ = 0$

As $\sin 60^\circ = \frac{\sqrt{3}}{2}$ and $\sin 30^\circ = \frac{1}{2}$ (see Book C2, Chapter 8):

$$P + 8 \times \frac{\sqrt{3}}{2} - S \times \tfrac{1}{2} = 0$$

$$P = \tfrac{1}{2}S - 4\sqrt{3}$$

Substitute the value calculated above for S:

$$P = \frac{14\sqrt{3}}{3} - 4\sqrt{3}$$

$$= \frac{2\sqrt{3}}{3}$$

So the force $\mathbf{P}$ has magnitude $\frac{2\sqrt{3}}{3}$ N (1.15 N) and the force $\mathbf{S}$ has magnitude $\frac{28\sqrt{3}}{3}$ N (16.2 N).

Example 8

A particle is in equilibrium under the forces $(6\mathbf{i} + 4\mathbf{j})$ N, $(-2\mathbf{i} - 5\mathbf{j})$ N and $(a\mathbf{i} + b\mathbf{j})$ N. Find the values of a and b.

As the particle is in equilibrium:

$$(6\mathbf{i} + 4\mathbf{j}) + (-2\mathbf{i} - 5\mathbf{j}) + (a\mathbf{i} + b\mathbf{j}) = \mathbf{0}$$

$$\mathbf{i}(6 - 2 + a) + \mathbf{j}(4 - 5 + b) = \mathbf{0}$$

As the resultant is the zero vector, the coefficient of $\mathbf{i}$ is zero. So:

$$6 - 2 + a = 0$$

$$a = -4$$

The coefficient of $\mathbf{j}$ is also zero. So:

$$4 - 5 + b = 0$$

$$b = 1$$

Exercise 4C

1

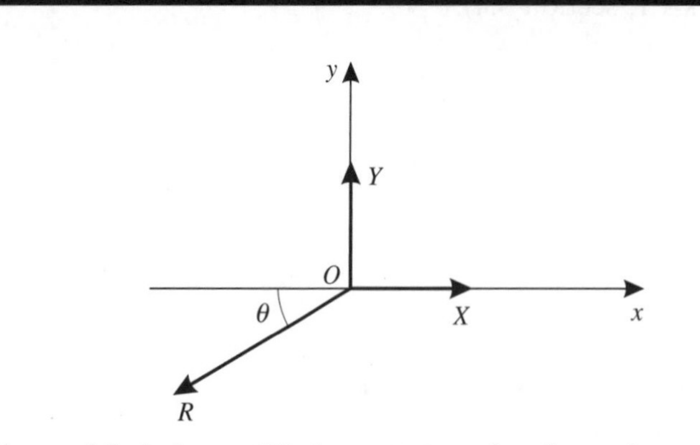

A particle is in equilibrium at O under the action of forces $\mathbf{X}$, $\mathbf{Y}$ and $\mathbf{R}$, as shown in the diagram. Find the magnitude of $\mathbf{R}$ and the angle θ when:

(a) $X = 3\,\text{N}$, $Y = 4\,\text{N}$

(b) $X = 5\,\text{N}$, $Y = 8\,\text{N}$

(c) $X = 5\,\text{N}$, $Y = 12\,\text{N}$.

2

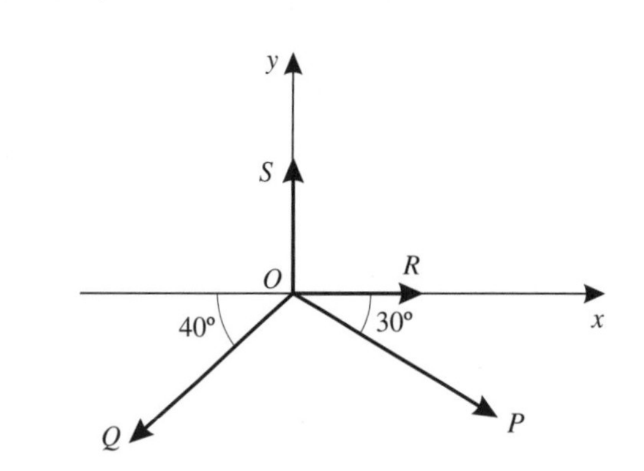

A particle is in equilibrium at O under the action of forces P, Q, R and S, as shown in the diagram. Find the magnitudes of forces R and S when:

(a) $P = 5\,\text{N}$, $Q = 6\,\text{N}$

(b) $P = 8\,\text{N}$, $Q = 10\,\text{N}$

(c) $P = 12\,\text{N}$, $Q = 20\,\text{N}$.

3 A particle is in equilibrium at O under the action of forces $\mathbf{P}\,\text{N}$, $\mathbf{Q}\,\text{N}$ and a third force $(x\mathbf{i} + y\mathbf{j})\,\text{N}$. Find x and y when:

(a) $\mathbf{P} = 2\mathbf{i} + 3\mathbf{j}$, $\mathbf{Q} = 3\mathbf{i} + 5\mathbf{j}$

(b) $\mathbf{P} = 2\mathbf{i} - 3\mathbf{j}$, $\mathbf{Q} = -3\mathbf{i} - 4\mathbf{j}$

(c) $\mathbf{P} = 3\mathbf{i} - \mathbf{j}$, $\mathbf{Q} = 2\mathbf{i} - 5\mathbf{j}$

(d) $\mathbf{P} = -4\mathbf{i} + 5\mathbf{j}$, $\mathbf{Q} = -\mathbf{i} - 7\mathbf{j}$

(e) $\mathbf{P} = -\mathbf{j}$, $\mathbf{Q} = \mathbf{i} + 2\mathbf{j}$.

4 Three forces $\mathbf{F}_1$, $\mathbf{F}_2$ and $\mathbf{F}_3$ act on a particle.
$\mathbf{F}_1 = (-6\mathbf{i} + 2\mathbf{j})$ N, $\mathbf{F}_2 = (p\mathbf{i} + 4\mathbf{j})$ N, $\mathbf{F}_3 = (4\mathbf{i} + q\mathbf{j})$ N. Given that the particle is in equilibrium, determine the values of p and q.

4.5 Types of force

So far the forces acting on a particle have been examined, without considering the *nature* of these forces. Some of the mechanical forces that may act on a particle are considered in this section.

Weight

All particles falling freely under gravity have the same acceleration. This constant acceleration is denoted by the symbol g. The acceleration must be caused by a force acting on the particle. This force is called the **weight** of the particle. If the mass of the particle is m kg then (as shown in section 5.1):

$$\text{weight} = mg$$

where the weight is a force measured in newtons.

Throughout this book the value of g will be taken as $9.8\,\text{m}\,\text{s}^{-2}$, although it has different values at different points on the surface of the Earth. Weight is an example of a **non-contact force**: an object does not have to be in contact with the Earth for the force of gravity to act on it. All the other forces considered below are **contact forces**.

Tension

Imagine a particle of mass m kg hanging in equilibrium at the end of a string. As the particle is in equilibrium (it doesn't fall or fly upwards) the weight mg acting downwards must be balanced by an equal and opposite force acting upwards. This force is called the **tension** T in the string.

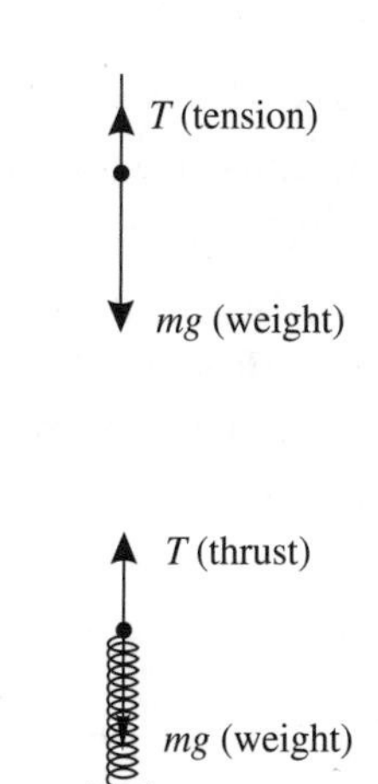

Thrust

Now imagine a similar situation where a particle is supported by a vertical spring from below. As the particle is in equilibrium there must be an upward force in the spring to balance the downward force of weight. This is called the **thrust**.

4.6 Resolving a contact force into normal and frictional components

Think of a particle at rest on a horizontal table. If the table is removed the particle will fall to the ground. There must therefore be an upward force, supplied by the table, which balances the weight of the particle. This is called the **normal reaction** or **normal contact force** and is usually denoted by **R**.

If you try to push the particle across the table then in general the particle will not move. An additional **frictional force F** acts to balance the pushing force. The frictional force acts parallel to the table and *opposes* the motion that the pushing force is trying to create. Here is the force diagram of this situation:

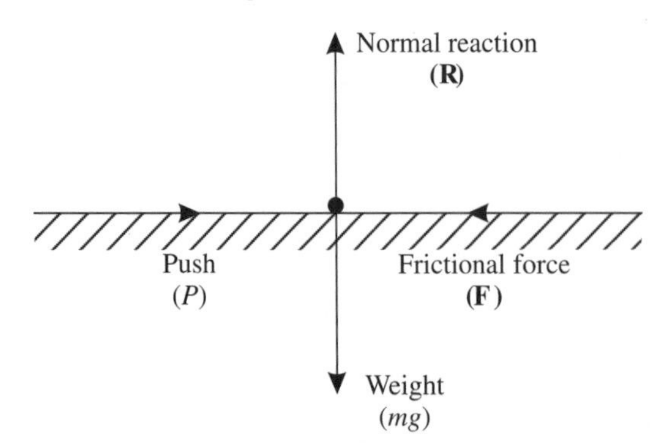

If the table has a perfectly *smooth* surface then $F = 0$ and there is only a normal reaction.

When you have a problem in which some or all of the above forces act on a particle, it is important to begin by drawing a clear diagram of reasonable size on which the forces are clearly indicated.

Example 9

Draw diagrams to show the forces acting on a particle in each of the following.

(a) A particle at rest on a *rough* inclined plane: **R** is perpendicular to the plane. **F** is up the plane as it opposes the motion which the weight of the particle on its own would produce.

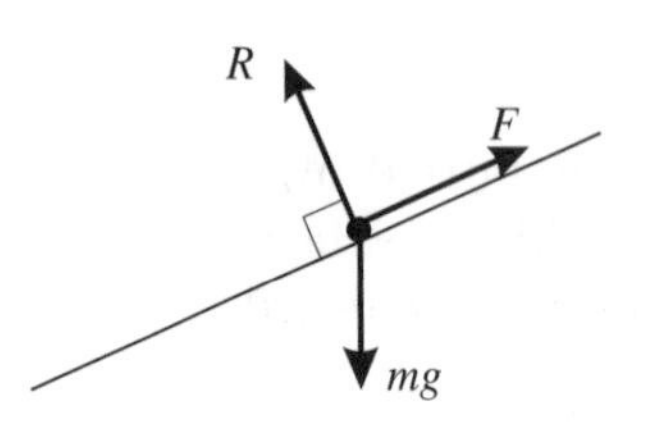

(b) A particle pulled across a *rough* horizontal table by a string inclined at 60° to the horizontal:

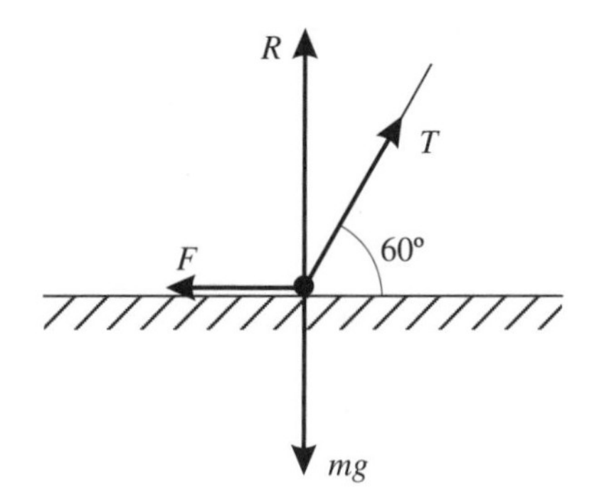

F opposes the force produced by the pull on the string.
(c) A particle held at rest on a *smooth* plane inclined at 30° to the horizontal, by a string parallel to the plane:

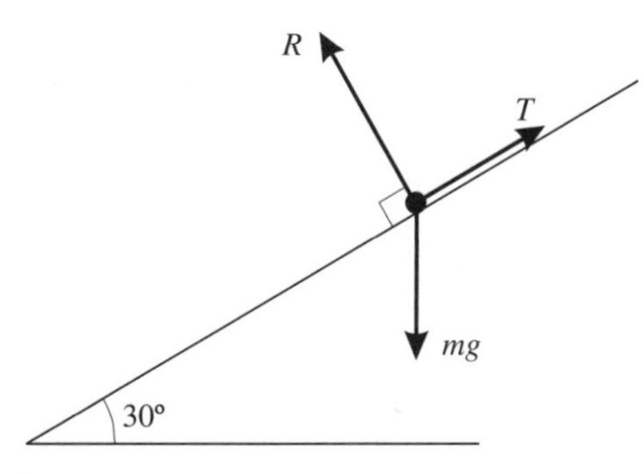

There is no frictional force in this case as the plane is smooth.
(d) A particle suspended from a horizontal beam by two unequal strings:

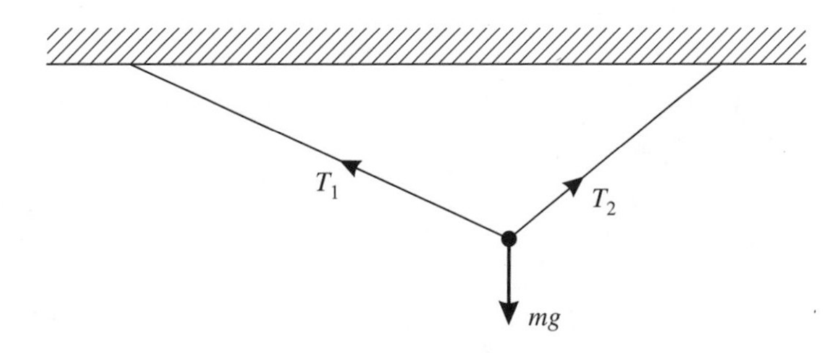

Some examples of equilibrium under these forces

Example 10

A particle of mass 5 kg is suspended in equilibrium by two light inextensible strings which make angles of 30° and 45° respectively with the horizontal. Find the tensions in the strings.

You will remember that 'inextensible' means the strings do not stretch.

The weight of the particle is $5g$ N (see p. 77). The force diagram is:

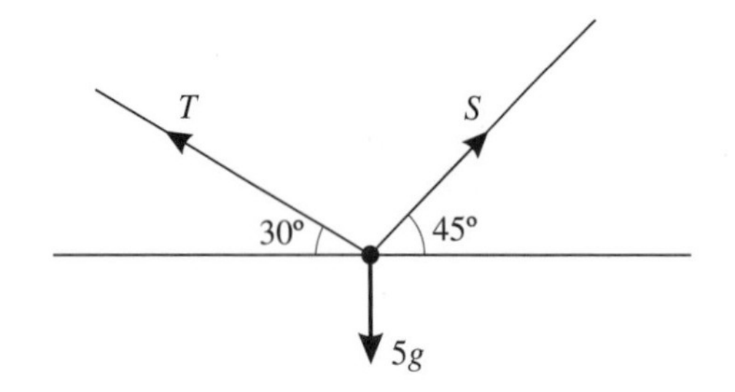

As the particle is in equilibrium, the components of the forces in each direction must add up to zero.

Resolving horizontally $\rightarrow$ gives: $S\cos 45^\circ - T\cos 30^\circ = 0$

$$T = \frac{S\cos 45^\circ}{\cos 30^\circ}$$

You may wish to keep the equation in trigonometrical form or you could use $\cos 45^\circ = \frac{\sqrt{2}}{2}$ and $\cos 30^\circ = \frac{\sqrt{3}}{2}$. (In the latter case you would get $T = S\sqrt{\frac{2}{3}}$.)

Resolving vertically $\uparrow$ gives: $T\sin 30^\circ + S\sin 45^\circ - 5g = 0$

Now substitute the value of T found above:

$$S\cos 45^\circ \times \left(\frac{\sin 30^\circ}{\cos 30^\circ}\right) + S\sin 45^\circ = 5g$$

$$S\cos 45^\circ \tan 30^\circ + S\sin 45^\circ = 5g$$

(or $S\frac{\sqrt{2}}{2}\left[\frac{1}{\sqrt{3}} + 1\right] = 5g$)

Using $g = 9.8$ gives:

$$S = 43.9\text{ N}$$

and: $$T = 35.9\text{ N}$$

Example 11

A particle of mass m kg rests in equilibrium on a rough plane inclined at 30° to the horizontal. Find the normal contact force and the frictional force in terms of m and g.

The force diagram is:

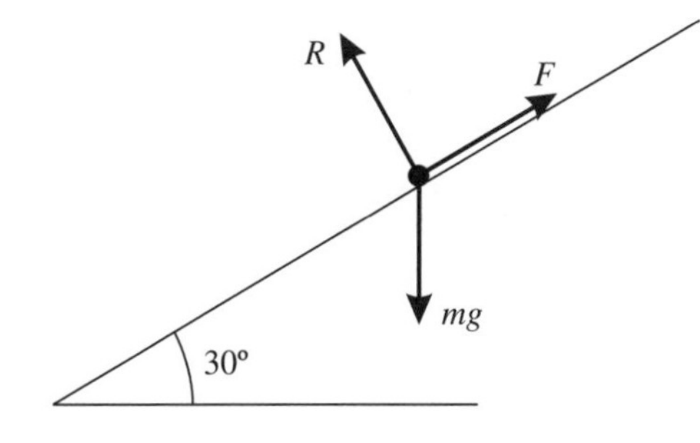

So that essentially:

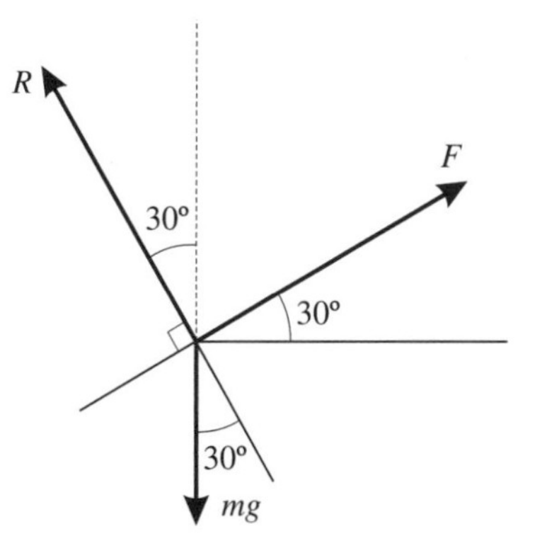

You can solve this problem in two ways, by choosing two different sets of directions in which to resolve the forces.

(i) Take components along and perpendicular to the plane. **F** is along the plane and **R** is perpendicular to the plane. You only need, therefore, to resolve the weight into components along and perpendicular to the plane.

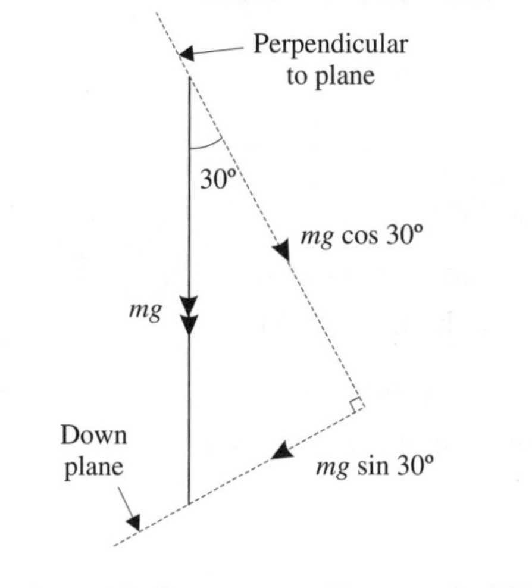

Resolving up the plane ↗ gives: $F - mg \sin 30^\circ = 0$

As $\sin 30^\circ = \frac{1}{2}$: $F = \dfrac{mg}{2}$

Resolving perpendicular to the plane ↖ gives:
$R - mg\ \cos 30° = 0$

As $\cos 30° = \frac{\sqrt{3}}{2}$: $\qquad R = \frac{mg\sqrt{3}}{2}$

(ii) Take components horizontally and vertically, using sin 30° and cos 30° as above:

Resolving vertically ↑ gives: $\qquad R \cos 30° + F \sin 30° = mg$

$$R \times \frac{\sqrt{3}}{2} + F \times \frac{1}{2} = mg$$

Resolving horizontally → gives: $\quad F \cos 30° - R \sin 30° = 0$

So: $\qquad R = F\sqrt{3}$

Substitute for R in the other equation to get:

$$F \times \tfrac{3}{2} + F \times \tfrac{1}{2} = mg$$

$$F = \frac{mg}{2}$$

Hence: $\qquad R = \frac{mg\sqrt{3}}{2}$

In the first method you obtained F and R directly. You may, however, find the equations easier to write down in the second method. Choose whichever method you find easier.

Exercise 4D

1 A small body is hanging in equilibrium at the end of a vertical string. Find

(a) the weight of the body if its mass is 0.2 kg

(b) the mass of the body if its weight is 4.9 N

(c) the tension in the string if the mass of the body is 1 kg

(d) the tension in the string if the weight of the body is 5 N.

2 A book is at rest in equilibrium on a horizontal table. Find

(a) the normal reaction of the table on the book if the weight of the book is 8 N

(b) the normal reaction of the table on the book if the mass of the book is 2.2 kg

(c) the weight of the book if the normal reaction is 6 N

(d) the mass of the book if the normal reaction is 10 N.

3

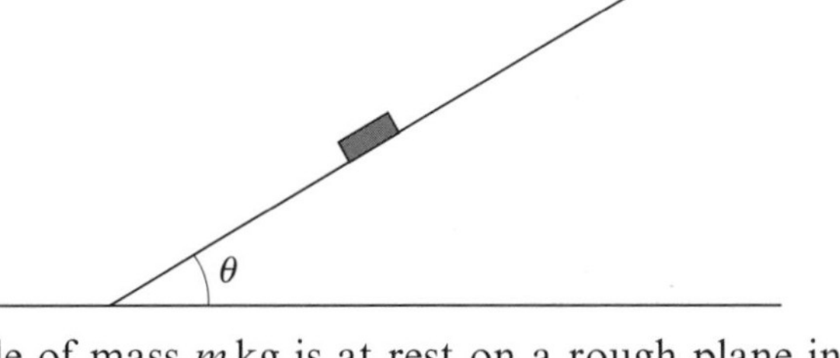

A particle of mass m kg is at rest on a rough plane inclined at an angle θ above the horizontal, as shown in the diagram. Find the frictional force exerted by the surface of the plane on the particle when:

(a) $m = 5$, $\theta = 30°$

(b) $m = 10$, $\theta = 40°$

(c) the weight of the particle is 8 N and $\theta = 20°$.

4

A particle of mass m kg is at rest on a rough plane inclined at an angle 30° above the horizontal. Force X N acts on the particle up the line of greatest slope of the plane, as shown in the diagram. Find the magnitude and direction of the frictional force when:

(a) $X = 10$, $m = 1$

(b) $X = 10$, $m = 2$

(c) $X = 10$, $m = 3$

(d) $X = 5$, $m = 2$

(e) $X = 5$ and the weight of the particle is 4 N.

5

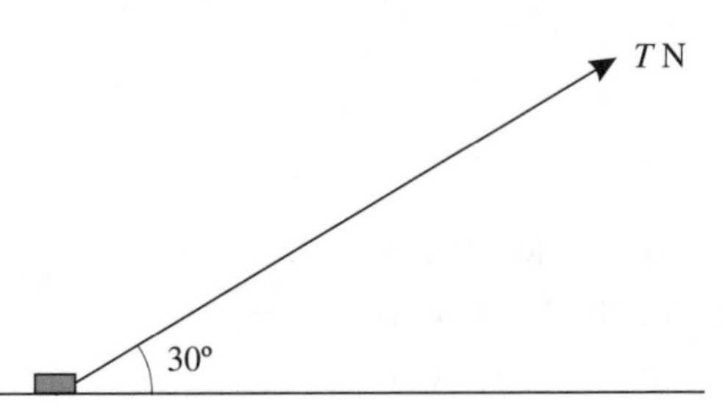

A small box of mass m kg is to be pulled across a rough horizontal floor at a constant speed by a string inclined at 30° above the horizontal, as shown in the diagram. The tension in the string is T N. Model the box as a particle and find the frictional force F N exerted by the floor on the box and the normal reaction of the floor on the box when the box is on the point of sliding when:

(a) $m = 5,\ T = 20$

(b) $m = 1,\ T = 10$

(c) $T = 20$ and the weight of the box is 12 N.

6 'Newton's cradle' consists usually of five small steel spheres, each suspended by two light strings. Modelling this by a number of particles each suspended by two strings inclined at 80° to the horizontal, find the tension in each string when:

(a) the weight of each steel sphere is 0.8 N

(b) the mass of each steel sphere is 0.1 kg.

7

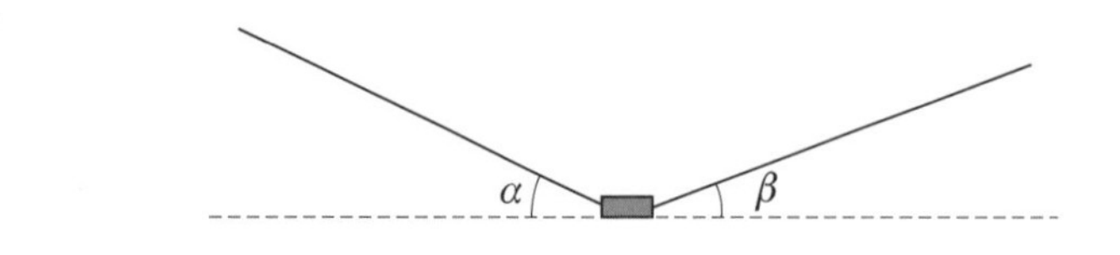

A peg-bag of mass m hanging from a washing line may be modelled by a particle supported in equilibrium by two straight strings inclined at angles α and β to the horizontal, as shown in the diagram. Find the tension in each string when:

(a) $m = 0.5,\ \alpha = \beta = 10°$

(b) $m = 0.5,\ \alpha = 10°,\ \beta = 12°$

(c) the weight of the bag is 2 N, $\alpha = 6°, \beta = 7°$.

Why in practice will α and β never differ by very much?

8 

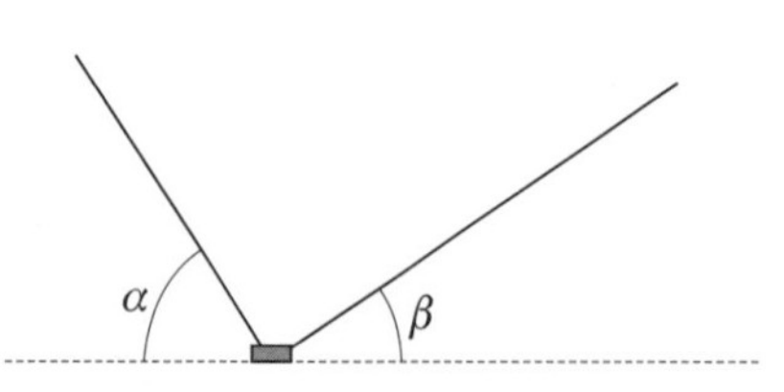

An electric light mass m kg in a workshop is held over a particular place by two strings attached to the light bulb holder. This is modelled by a particle suspended by two straight strings inclined at angles α and β above the horizontal. Find the tension in the strings when:

(a) $m = 0.5,\ \alpha = \beta = 60°$

(b) $m = 0.5,\ \alpha = 60°,\ \beta = 65°$.

4.7 Friction and the coefficient of friction

Now have a closer look at the frictional force described in section 4.6.

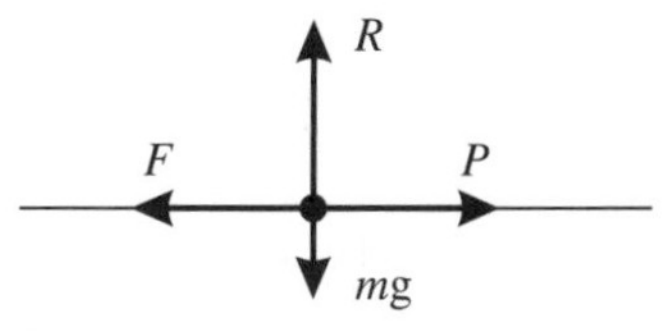

Remember that if a horizontal force **P** is applied to a particle lying on a rough horizontal table, the particle does not necessarily move. There is a frictional force **F** opposing the force **P**.

If **P** and **F** are the only horizontal forces acting on the particle then so long as the particle is stationary:

$$P = F$$

- **The magnitude of the frictional force is just sufficient to prevent relative motion.**

The frictional force **F** for a particular surface is not constant. It increases as the applied force **P** increases until the force **F** reaches a value $\mathbf{F}_{max}$ beyond which it cannot increase. The particle is then just about to move and is said to be in a state of **limiting equilibrium**. At this point friction is said to be **limiting**.

It can be shown experimentally that F_{max} is proportional to R, that is:

$$F_{max} = \mu R$$

The constant of proportionality, which is always given as the symbol μ, is called the **coefficient of friction**. (μ is a letter of the Greek alphabet and is pronounced 'mu'.)

In general, therefore, $F \leqslant \mu R$. Clearly $F \geqslant 0$ and is only zero when the surface is completely smooth. (In real life, no surface *is* completely smooth!)

Once the particle begins to move the frictional force opposing the relative motion *remains* at the constant value μR. The value of μ depends on the nature of the two surfaces in contact.

- To summarise:

  1. **Friction acts to oppose relative motion.**
  2. **Until it reaches its limiting value the magnitude of the frictional force is just sufficient to prevent relative motion.**
  3. **When the limiting value is reached $F = \mu R$, where R is the normal contact force and μ is the coefficient of friction.**

4. **For all *rough* surfaces $0 < F \leqslant \mu R$.**
5. **For a *smooth* surface $F = 0$.**
6. **When the particle begins to slide *the frictional force takes its limiting value* μR and acts in the direction *opposite* to the direction of relative motion.**

Example 12

A particle of weight 30 N rests on a horizontal plane. The coefficient of friction between the particle and the plane is 0.3. A horizontal force of magnitude P N is applied to the particle. Find the value of P given that the particle is about to slide.

The force diagram is:

R N

F N P N

30 N

Resolving horizontally $\rightarrow$ gives:
$$P - F = 0$$
$$P = F$$

Resolving vertically $\uparrow$ gives:
$$R - 30 = 0$$
$$R = 30$$

When the particle is about to slide, friction is limiting, that is:

$$F = \mu R$$

So: $P = F = \mu R = (0.3) \times 30$

Hence: $P = 9$

Example 13

A particle of weight 30 N rests in equilibrium on a rough horizontal table. A string is attached to the particle. The string makes an angle of 30° with the horizontal and the tension in the string is 18 N. Find the magnitude of the frictional force acting on the particle.

The forces acting on the particle are the tension T along the string and the weight 30 N. As a result of the contact with the table there will be a normal reaction R N and a frictional force F N parallel to the table. The direction of F N will be such as to oppose any subsequent motion. The force diagram is:

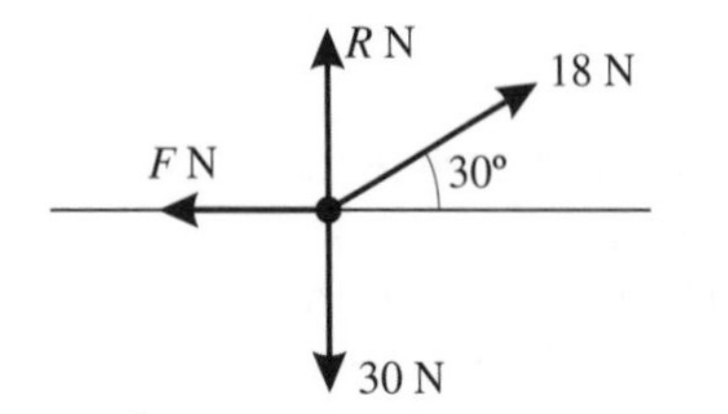

You want to find F, so resolve horizontally:

$$\rightarrow \quad 18\cos 30^\circ - F = 0$$

$$F = 18\cos 30^\circ$$

$$= 18\left(\frac{\sqrt{3}}{2}\right) = 9\sqrt{3} = 15.6$$

The magnitude of the frictional force is 15.6 N.

Example 14

A particle of mass 4 kg rests in limiting equilibrium on a rough plane inclined at 30° to the horizontal. Find the coefficient of friction between the particle and the plane.

The force diagram is:

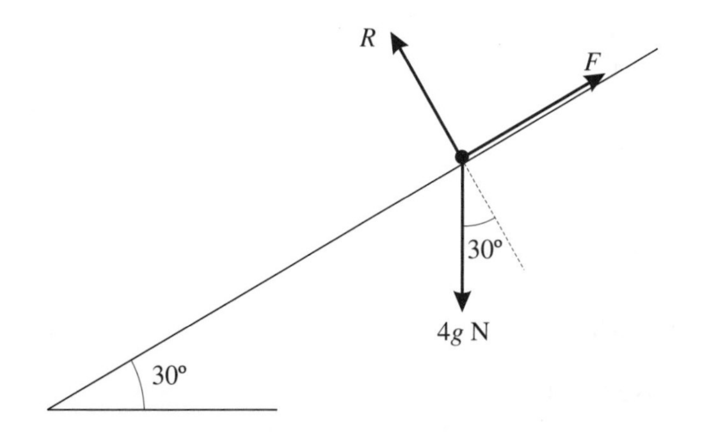

Notice that **F** acts *up* the plane, since any subsequent motion will be *down* the plane. (Things slide downhill, not uphill!)

Resolving perpendicular to the plane ↖ gives:

$$R - 4g\cos 30^\circ = 0$$

$$R = 4g\cos 30^\circ = 4g \times \frac{\sqrt{3}}{2} = 2g\sqrt{3}$$

Resolving up the plane ↗ gives: $F - 4g\sin 30^\circ = 0$

$$F = 4g\sin 30^\circ = 4g \times \tfrac{1}{2} = 2g$$

As equilibrium is limiting, $F = \mu R$ and so:

$$2g = \mu \times 2g\sqrt{3}$$

$$\mu = \frac{1}{\sqrt{3}} = 0.577$$

The coefficient of friction between the particle and the plane is 0.577.

Example 15

A parcel of mass 3 kg rests in limiting equilibrium on a rough plane inclined at 30° to the horizontal. The coefficient of friction between the parcel and the plane is $\frac{1}{3}$. A horizontal force of magnitude X N is applied to the particle so that equilibrium will be broken by the particle moving upwards. Find the value of X.

The first step in the solution of this problem is to replace the parcel by a particle of mass 3 kg. The force diagram is then:

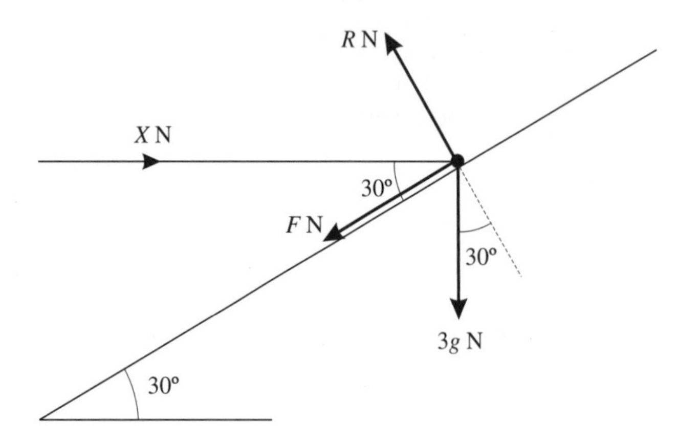

Notice that **F** is down the plane as you are told that subsequent motion will be *up* the plane.

Resolving perpendicular to the plane ↖ gives:

$$R - X\sin 30° - 3g\cos 30° = 0$$

So:

$$R - X \times \tfrac{1}{2} - 3g \times \frac{\sqrt{3}}{2} = 0 \qquad (1)$$

Resolving down the plane ↙ gives:

$$F + 3g\sin 30° - X\cos 30° = 0$$

So:

$$F + 3g \times \tfrac{1}{2} - X \times \frac{\sqrt{3}}{2} = 0 \qquad (2)$$

As the equilibrium is limiting:

$$F = \mu R = \tfrac{1}{3}R$$

Substituting this result for F in equation (2) gives:

$$\tfrac{1}{3}R + \frac{3g}{2} - X\frac{\sqrt{3}}{2} = 0$$

Substituting for R from equation (1) gives:

$$\tfrac{1}{3}\left(\tfrac{1}{2}X + 3g\frac{\sqrt{3}}{2}\right) + \frac{3g}{2} - X\frac{\sqrt{3}}{2} = 0$$

So:
$$X = \frac{g(9+3\sqrt{3})}{3\sqrt{3}-1}$$
$$= \frac{9.8(9+3\sqrt{3})}{3\sqrt{3}-1}$$
$$= 33.2$$

The value of X is 33.2.

Exercise 4E

Whenever a numerical value of g is required, take $g = 9.8\,\text{m}\,\text{s}^{-2}$.

1 A particle of weight mg is at rest on a rough horizontal plane. The coefficient of friction between the particle and the plane is μ. A horizontal force **P** is applied to the particle, which is now about to slip. Find
(a) P, given $mg = 5\,\text{N}$, $\mu = 0.4$
(b) μ given $mg = 4\,\text{N}$, $P = 2\,\text{N}$
(c) mg, given $\mu = 0.3$, $P = 6\,\text{N}$.

2 A particle is in equilibrium on a rough horizontal table. A string is attached to the particle and held at an angle 30° to the horizontal. The tension in the string is T, the frictional force exerted by the table on the particle is F. Find
(a) F, given $T = 25\,\text{N}$, (b) T, given $F = 25\,\text{N}$.

3 The particle in question 2 is about to slip on the table. Given that the mass of the particle is 5 kg, find for each part of the question the normal reaction between the particle and the table and the coefficient of friction.

4 A particle of mass 2 kg rests in limiting equilibrium on a rough plane inclined at 20° to the horizontal. Find
(a) the frictional force exerted by the plane on the particle,
(b) the coefficient of friction between the particle and the plane.

5 A body of mass 2 kg is held in limiting equilibrium on a rough plane inclined at 20° to the horizontal by a horizontal force **X**. The coefficient of friction between the body and the plane is 0.2. Modelling the body as a particle find X when the body is on the point of slipping
(a) up the plane, (b) down the plane.

6 The force X in Question 5 is replaced by a force $\mathbf{Y}$ at an angle of 45° to the horizontal. Find Y when the body is on the point of slipping

(a) up the plane,

(b) down the plane.

SUMMARY OF KEY POINTS

1. Force is a vector quantity.
2. Forces can be added by using the triangle law or parallelogram rule.
3. The resultant of a system of forces is most easily found by using components.
4. A system of forces is in equilibrium if their lines of action pass through a single point and if their resultant is the zero vector.
5. The magnitude of the frictional force is just sufficient to prevent relative motion.
6. For a smooth surface there is no frictional force ($F = 0$).
7. When sliding occurs the frictional force takes its limiting value μR and opposes the relative motion

Dynamics of a particle moving in a straight line or plane

5

What is dynamics?

Chapter 3 of this book considers the motion of a particle without any reference to how that motion was produced. Chapter 4 looks at the forces which act on a particle without reference to motion. In many cases the forces acting on a particle will cause motion. **Dynamics** is the study of the relationship between the forces acting on a moving particle and the motion of the particle.

5.1 Newton's laws of motion

Newton's first law

A particle experiences a gravitational force called its weight. If this is the only force acting on the particle it will fall to earth, as seen in chapter 3. If however the particle is resting on a surface, that surface will exert a force on it preventing it from falling. This contact force will balance the weight of the particle and no motion will take place. If the particle is initially at rest on a rough surface and is acted on by a force with a component parallel to the surface of sufficient magnitude to overcome any friction between the particle and the surface, the particle will move along the surface. Usually the particle will be seen to gain speed – in other words it will have an acceleration. A particle moving at constant speed has no acceleration and there is no resultant force acting on it.

- **A particle will remain at rest or will continue to move with constant velocity in a straight line unless acted on by a resultant force.**

Newton's second law

A resultant force acting on a particle will cause the particle to accelerate. The acceleration is proportional to the force producing it. The same force applied to particles of different mass will produce different accelerations. The force needed to produce a given acceleration is proportional to the mass of the body being accelerated.

Mass is measured in kilograms (kg) and the basic unit of force is the newton (N). The **newton** is defined so that a force of 1 newton produces an acceleration of $1\,\mathrm{m\,s^{-2}}$ when applied to a particle of mass 1 kg.

A force of **F** newtons applied to a particle of mass m kg will result in an acceleration of $\mathbf{a}\,\mathrm{m\,s^{-2}}$ where:

$$\mathbf{F} = m\mathbf{a}$$

Force and acceleration are vectors because each has a magnitude and a direction. Therefore the acceleration produced is in the direction of the resultant force acting on the body.

Newton's second law can be summarised by the equation $\mathbf{F} = m\mathbf{a}$. This is often called the **equation of motion** of the particle.

- **The force F applied to a particle is proportional to the mass m of the particle and the acceleration produced.**

$$\mathbf{F} = m\mathbf{a}$$

Newton's third law

- **Every action has an equal and opposite reaction.**

This means that if a particle A exerts a force on a particle B then B exerts a force on A of the same magnitude but in the opposite direction. These equal and opposite forces are known as **reactions** between the particles. Newton's third law is also true when two bodies which are not both particles have contact. This law is used when studying normal contact forces – as in chapter 4. A particle at rest on the floor exerts a force on that floor. The floor also exerts a force of equal magnitude but opposite direction on the particle and the particle remains at rest.

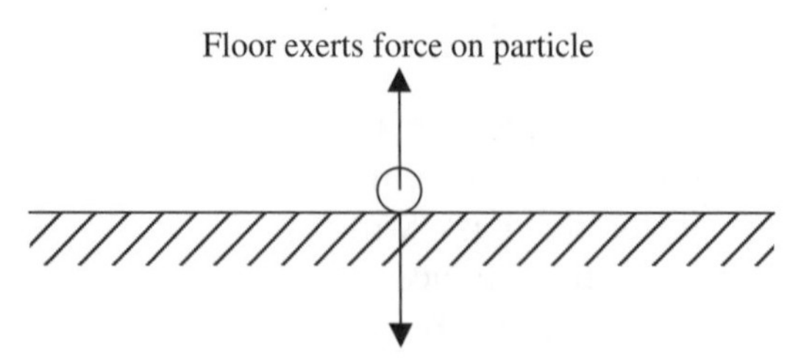

Example 1

Find the magnitude of the acceleration produced when a particle of mass 4 kg is acted on by a resultant force of magnitude 16 N.

The acceleration is parallel to the resultant force as shown in the diagram:

$a\ \mathrm{m\,s^{-2}}$

4 kg ○ → 16 N

Let the acceleration be $a\ \mathrm{m\,s^{-2}}$.

Using: $F = ma$

Gives: $16 = 4 \times a$

$a = 4$

The magnitude of the acceleration is $4\ \mathrm{m\,s^{-2}}$.

Example 2

A particle of mass 3 kg is accelerating at $12\ \mathrm{m\,s^{-2}}$. Find the magnitude of the resultant force on the particle.

The resultant force is parallel to the acceleration.

$12\ \mathrm{m\,s^{-2}}$

3 kg ○ → F N

Let the resultant force be FN.

Using: $F = ma$

Gives: $F = 3 \times 12 = 36$

The resultant force is of magnitude 36 N.

Example 3

Find in vector form the acceleration of a particle of mass 600 g when forces of $(7\mathbf{i} + 13\mathbf{j})$ N, $(4\mathbf{i} + 4\mathbf{j})$ N and $(-2\mathbf{i} - 5\mathbf{j})$ N act on it. Find also the magnitude and direction of the acceleration.

You must first convert the mass of the particle to kilograms.

Mass of particle $= 600\ \mathrm{g} = 0.6\ \mathrm{kg}$, as $1\ \mathrm{kg} = 1000\ \mathrm{g}$.

Resultant force $= (7\mathbf{i} + 13\mathbf{j}) + (4\mathbf{i} + 4\mathbf{j}) + (-2\mathbf{i} - 5\mathbf{j})$

$= 9\mathbf{i} + 12\mathbf{j}$

Using: $\mathbf{F} = m\,\mathbf{a}$

Gives: $9\mathbf{i} + 12\mathbf{j} = 0.6\,\mathbf{a}$

$$\mathbf{a} = \frac{9\mathbf{i} + 12\mathbf{j}}{0.6}$$
$$= 15\mathbf{i} + 20\mathbf{j}$$

The acceleration is $(15\mathbf{i} + 20\mathbf{j})\,\mathrm{m\,s^{-2}}$.

The magnitude of **a** is:

$$|\mathbf{a}| = \sqrt{(15^2 + 20^2)} = 25$$

Also:
$$\tan\theta = \frac{20}{15}$$
$$\theta = 53.1^\circ$$

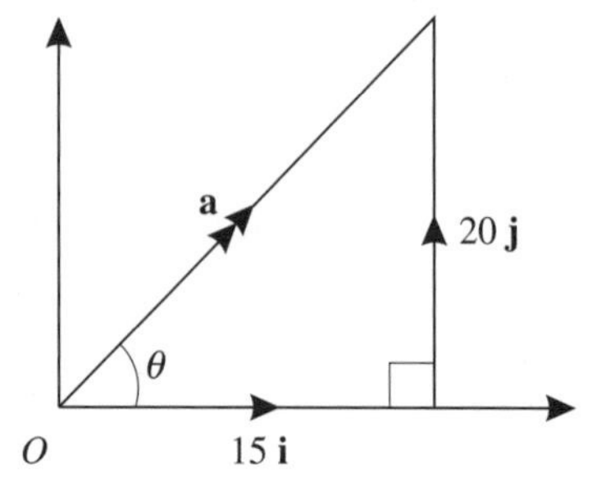

The acceleration has magnitude $25\,\mathrm{m\,s^{-2}}$ and is at an angle of 53.1° to the vector **i**.

Example 4

A car travels a distance of 32 m along a straight road while uniformly accelerating from rest to $16\,\mathrm{m\,s^{-1}}$. By modelling the car as a particle, find the acceleration of the car. Given that the mass of the car is 640 kg, find the magnitude of the accelerating force.

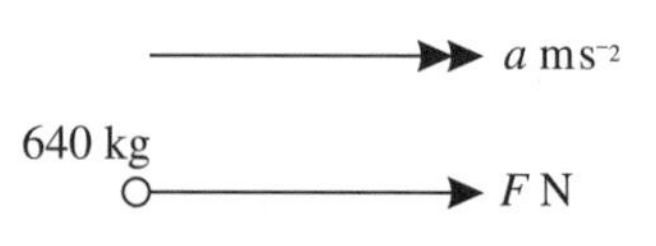

First apply the uniform acceleration formulae to find the acceleration of the car.

Known quantities are:

$$s = 32\ \mathrm{m}$$
$$u = 0\,\mathrm{m\,s^{-1}}$$
$$v = 16\,\mathrm{m\,s^{-1}}$$

Using: $v^2 = u^2 + 2as$

Substituting for v, u and s gives:

$$16^2 = 0^2 + 2 \times a \times 32$$
$$a = \frac{16^2}{2 \times 32} = 4$$

The acceleration of the car is $4\,\mathrm{m\,s^{-2}}$.

To find the magnitude of the accelerating force you must use the equation of motion for the car.

$$F = ma$$
$$F = 640 \times 4 = 2560$$

The accelerating force has magnitude 2560 N.

Example 5

A car of mass 600 kg is travelling along a straight horizontal road with an acceleration of $2\,\text{m s}^{-2}$. The engine is exerting a forward force of magnitude 1500 N. By modelling the car as a particle, find the magnitude of the resistance it is experiencing.

Let the resistance be R newtons, as shown in the diagram.
The resultant force on the car in the direction of motion is $(1500 - R)$N.

Using: $F = ma$

Substituting for F gives:

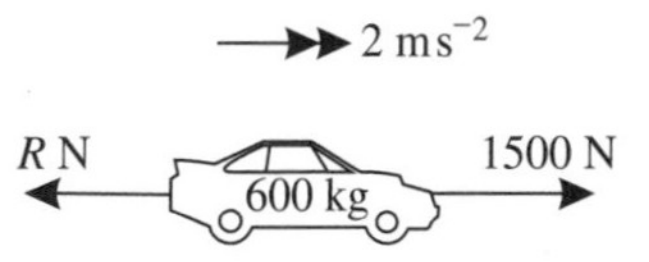

$$1500 - R = 600 \times 2$$

$$R = 1500 - 1200 = 300$$

The magnitude of the resistance is 300 N.

Exercise 5A

1 Find the acceleration produced when a particle of mass 2 kg is acted on by a resultant force of 12 N.

2 Find the resultant force which will produce an acceleration of $5\,\text{m s}^{-2}$ for a particle of mass 6 kg.

3 A particle of mass 5 kg has an acceleration of $4\,\text{m s}^{-1}$. Find the magnitude of the resultant force acting on the particle.

4 A particle of mass 4 kg experiences a resultant force of 20 N. Find the acceleration produced.

5 Find the resultant force which will produce an acceleration of $2\,\text{m s}^{-2}$ for a particle of mass 5 kg.

6 A particle has an acceleration of $3\,\text{m s}^{-2}$ when acted on by a resultant force of 12 N. Find the mass of the particle.

7 A particle of mass 5 kg is acted on by a force of $(4\mathbf{i} + 6\mathbf{j})$ N. Find in vector form the resulting acceleration.

8 A particle of mass 2 kg is accelerating at $(4\mathbf{i} - 2\mathbf{j})\,\text{m s}^{-2}$. Find in vector form the resultant force acting on the particle.

9 A particle of mass 2 kg is acted on by forces of $(5\mathbf{i} - 2\mathbf{j})$ N and $(3\mathbf{i} + 4\mathbf{j})$ N. Find in vector form the acceleration produced.

10 A packing case of mass 60 kg is dragged across a rough horizontal floor by a horizontal force of 50 N. The acceleration of the case is $0.4\,\text{m}\,\text{s}^{-2}$. Model the packing case as a particle and hence find the magnitude of the friction force.

11 A car travels a distance of 30 m along a straight horizontal road while uniformly accelerating from rest to $15\,\text{m}\,\text{s}^{-1}$. Find the acceleration of the car. Given that the car has a mass of 560 kg find the magnitude of the accelerating force.

12 A toy dog of mass 0.5 kg is pulled across a smooth horizontal floor by a horizontal string attached to the dog. The tension in the string is 0.75 N. By modelling the dog as a particle calculate the acceleration produced. Given that the dog starts from rest, determine how far it will move in 2 seconds.

13 A car of mass 550 kg is travelling along a straight horizontal road with an acceleration of $2.2\,\text{m}\,\text{s}^{-2}$. The force exerted by the engine is 1600 N. Find the magnitude of the resistance to motion.

14 Forces of $(9\mathbf{i} + 3\mathbf{j})$ N, $(7\mathbf{i} + 3\mathbf{j})$ N and $(a\mathbf{i} + b\mathbf{j})$ N, where a and b are constants, act on a particle of mass 2 kg. Given that the acceleration produced is $(10\mathbf{i} + 2\mathbf{j})\,\text{m}\,\text{s}^{-2}$ find the values of a and b.

15 A lorry of mass 2.5 tonnes experiences resistances totalling 800 N when travelling along a level road. The engine is producing a driving force of 2400 N. By modelling the lorry as a particle find the acceleration of the lorry.

16 A car of mass 600 kg is brought to rest in 6 seconds from a speed of $20\,\text{m}\,\text{s}^{-1}$. Neglecting resistances, find the braking force required to achieve this.

17 A car of mass 600 kg experiences a resistive force of 500 N while being brought to rest in 6 seconds from a speed of $20\,\text{m}\,\text{s}^{-1}$. Calculate the braking force required to achieve this.

18 Given that the resistances total 400 N find the magnitude of the constant force needed to accelerate a car of mass 800 kg from rest to $20\,\text{m s}^{-1}$ in 15 s.

19 Find the magnitude of the resultant force required to give a particle of mass 3 kg an acceleration of $(2\mathbf{i} - 3\mathbf{j})\,\text{m s}^{-2}$.

20 A stone slides in a straight line across a frozen pond. Given that the initial speed of the stone is $5\,\text{m s}^{-1}$ and that it slides 20 m before coming to rest, calculate the coefficient of friction between the stone and the surface of the frozen pond.

Vertical motion

When a particle is moving vertically, one of the forces acting in the direction of motion is the weight of the particle. A particle falling freely under gravity has an acceleration of $g = 9.8\,\text{m s}^{-2}$. Because the only force acting on the particle is its weight, the equation of motion is:

■ $$\mathbf{F} = \textbf{mass} \times g$$

or: $$\textbf{weight} = mg$$

Example 6

A particle of mass 2 kg is attached to the lower end of a string hanging vertically. The particle is lowered and moves with an acceleration of $0.2\,\text{m s}^{-2}$. Find the tension in the string.

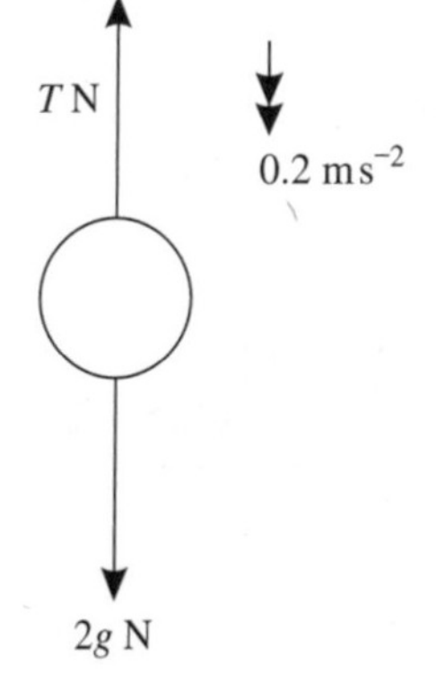

Mass of particle = 2 kg.

Therefore weight of particle = $2g$ N.

Let the tension in the string be T N as shown in the diagram.

Because the particle is moving downwards, the resultant force acting on the particle is downwards.

So the resultant downwards force $= (2g - T)$ N

Using: $F = ma$

Substituting for F gives: $2g - T = 2 \times 0.2$

$$T = 2 \times 9.8 - 2 \times 0.2 = 19.2$$

The tension in the string is 19.2 N.

Example 7

A stone of mass 0.5 kg is released from rest on the surface of the water in a well. It takes 2 seconds to reach the bottom of the well. Given that the water exerts a constant resistance of 2 N, find the depth of the well.

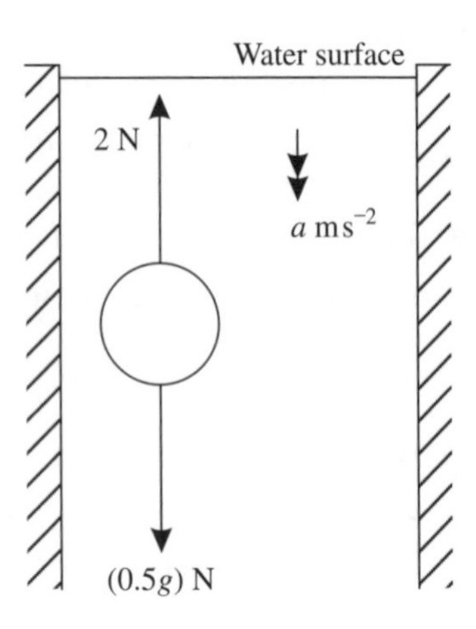

Using : $F = ma$

Gives : $0.5g - 2 = 0.5a$

$$a = \frac{0.5 \times 9.8 - 2}{0.5} = 5.8$$

The acceleration of the particle is $5.8\,\text{m s}^{-2}$.

The forces acting on the stone are constant throughout the motion. This means that the acceleration is also constant and the uniform acceleration equations can be applied.

Known quantities are:

$$u = 0\,\text{m s}^{-1}$$
$$a = 5.8\,\text{m s}^{-2}$$
$$t = 2\,\text{s}$$

Using: $s = ut + \frac{1}{2}at^2$

Substituting for u, a and b gives:

$$s = 0 + \tfrac{1}{2} \times 5.8 \times 2^2$$
$$= 11.6$$

The well is 11.6 m deep.

Example 8

A parcel of mass 5 kg is released from rest on a rough ramp of inclination $\theta = \arcsin \frac{3}{5}$ (that is $\sin\theta = \frac{3}{5}$) and slides down the ramp. After 3 seconds the parcel has a speed of $4.9\,\text{m s}^{-1}$. Treating the parcel as a particle, find the coefficient of friction between the parcel and the ramp. (See chapter 4 for work on the coefficient of friction.)

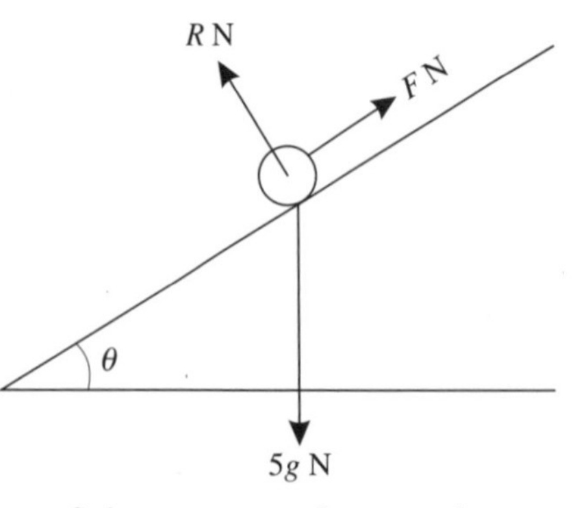

R N is the reaction of the ramp on the parcel.

The component of the weight perpendicular to the ramp is $5g \cos\theta$ N.

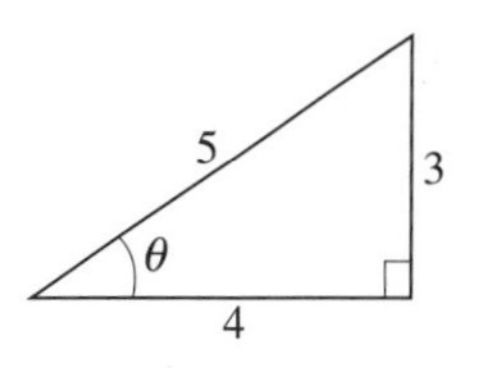

There is no motion perpendicular to the ramp so resolving perpendicular to the ramp gives:

$$R - 5g\cos\theta = 0$$

$$R = 5g\cos\theta = 5g \times \tfrac{4}{5} = 4g$$

The particle has an acceleration parallel to the ramp.

The friction force acts up the ramp to oppose the motion.

The component of weight down the ramp is $5g \sin\theta$ N. Using the equation of motion down the ramp gives:

$$5g\sin\theta - F = 5a \qquad (1)$$

Use the uniform acceleration equations to find a.

Known quantities are:

$$t = 3\,\text{s}$$

$$u = 0\,\text{m s}^{-1}$$

$$v = 4.9\,\text{m s}^{-1}$$

Using: $\quad v = u + at$

Substituting for t, u and v gives:

$$4.9 = 0 + 3a$$

$$a = \frac{4.9}{3}$$

Substituting this value for a in equation (1) and using $\sin\theta = \frac{3}{5}$ gives:

$$F = 5g \times \tfrac{3}{5} - 5 \times \frac{4.9}{3}$$

$$F = 3 \times 9.8 - \frac{5 \times 4.9}{3}$$

$F = 21.23$ and from above $R = 4g$

For sliding $F = \mu R$

$$\mu = \frac{F}{R}$$

Substituting $R = 4g$ and $F = 21.23$ gives:

$$\mu = \frac{21.23}{4 \times 9.8} = 0.5415$$

The coefficient of friction is 0.542.

Example 9

A lift is accelerating upwards at $1.5\,\text{m}\,\text{s}^{-2}$. A child of mass 30 kg is standing in the lift. Treating the child as a particle find the force between the child and the floor of the lift.

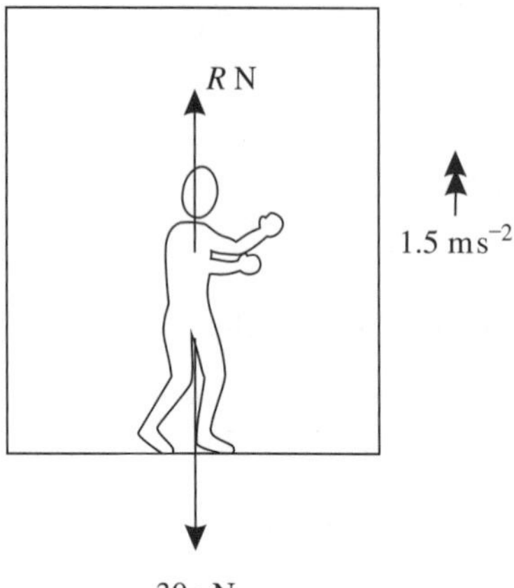

The child exerts a force on the floor of the lift.

By Newton's third law, the floor exerts an equal and opposite force on the child. Let this force be R N as shown in the diagram.

The resultant upwards force on the child is $(R - 30g)$ N.

Using: $F = ma$

Gives: $R - 30g = 30 \times 1.5$

$$R = 30 \times 1.5 + 30 \times 9.8$$

$$R = 339$$

The force between the child and the floor is 339 N.

Exercise 5B

Whenever a numerical value of g is required take $g = 9.8\,\text{m}\,\text{s}^{-2}$.

1 A particle of mass 0.3 kg is attached to the lower end of a vertical string. If the particle has a downwards acceleration of magnitude $0.9\,\text{m}\,\text{s}^{-2}$ find the tension in the string.

2 A stone of mass 0.06 kg is falling through a liquid with an acceleration of $3.6\,\text{m}\,\text{s}^{-2}$. Find the resistive force acting on the stone.

3 A boy of mass 50 kg travels up in a lift. The acceleration of the lift is $0.4\,\text{m}\,\text{s}^{-2}$. Find the force exerted on the boy by the floor of the lift. What force does the boy exert on the floor of the lift?

4 A cable raising a load with an acceleration of $1.5\,\text{m}\,\text{s}^{-2}$ has a tension of 15 kN. Find the mass of the load.

5 A boy is tobogganing down a smooth slope inclined at 20° to the horizontal. Find his acceleration, assuming resistances can be neglected.

6 A boy is tobogganing down a slope inclined at 25° to the horizontal. The resistances to his motion amount to 15 N. By modelling the boy and his toboggan as a single particle, find the mass of the boy and his toboggan when his acceleration is $3.9\,\text{m}\,\text{s}^{-2}$.

7 A particle of mass 9 kg is sliding down a smooth inclined plane with an acceleration of $4.9\,\text{m}\,\text{s}^{-2}$. Find the angle of inclination of the plane.

8 A particle of mass 6 kg is sliding down a smooth plane inclined at 45° to the horizontal. Find the acceleration of the particle.

9 A particle of mass 2 kg is sliding down a plane inclined at 24° to the horizontal against a resistive force of 3 N. Find the acceleration of the particle.

10 A parcel of mass 3 kg is sliding down a rough slope of inclination arcsin 0.4. The coefficient of friction between the parcel and the slope is 0.35. By modelling the parcel as a particle find its acceleration.

11 A parcel of mass 60 kg is released from rest on a smooth plane inclined at arcsin $\frac{3}{5}$ to the horizontal. Find the velocity of the parcel when it has travelled 5 m down the plane.

12 A particle of mass 2 kg slides down a rough plane inclined at 20° to the horizontal. Given that the acceleration of the particle is $1.5\,\text{m}\,\text{s}^{-2}$ find the coefficient of friction between the particle and the plane.

13 A particle of mass 4 kg is being pulled up a rough plane inclined at 25° to the horizontal by a force of magnitude 30 N acting along a line of greatest slope of the plane. Given that the particle is accelerating at $2\,\text{m}\,\text{s}^{-2}$ find the coefficient of friction between the particle and the plane.

14 A package of mass 10 kg is released from rest on a rough slope inclined at 25° to the horizontal. After 2 seconds the package has moved 4 m down the slope. Find the coefficient of friction between the package and the slope.

15 A horizontal force of 2 N is just sufficient to prevent a block of mass 1 kg from sliding down a rough plane inclined at arcsin $\frac{7}{25}$ to the horizontal. Find the coefficient of friction between the block and the plane and the acceleration with which the block will move when the force is removed.

16 A particle rests in **limiting equilibrium** (that is the particle is on the point of moving and friction has its maximum value) on a plane inclined at $30°$ to the horizontal. Determine the acceleration with which the particle will slide down the plane when the angle of inclination is increased to $40°$.

17 A particle of mass 2 kg rests in limiting equilibrium on a plane inclined at $25°$ to the horizontal. The angle of inclination is decreased to $20°$ and a force of magnitude 20 N is applied up a line of greatest slope. Find the particle's acceleration. When the particle has been moving for 2 seconds the force is removed. Determine the further distance the particle will move up the plane.

18 A block of mass 1.6 kg is placed on a rough plane inclined at $45°$ to the horizontal. The coefficient of friction between the block and the plane is $\frac{1}{4}$. Model the block as a particle and hence find the acceleration of the block down the plane. Find the velocity of the block after 2 seconds, assuming that it starts from rest.

5.2 The motion of two connected particles

When two moving particles are connected by a string which is light and inextensible there will be a tension in the string. By Newton's third law, the forces acting on the particles will have the same magnitude but will act in opposite directions as shown in the diagram.

T T

It is important on diagrams to show these separate tensions clearly.

It will be assumed that all strings are light and inextensible.

Example 10

Two particles A and B of masses 3 kg and 4 kg respectively connected by a light inextensible string are at rest on a smooth horizontal surface. A force of magnitude 7 N is applied to particle B in the direction AB. Find the acceleration produced and the tension in the string.

a m s^{-2}
T N
T N 4 kg
3 kg
7 N
A
B

The motion of each of the particles A and B needs to be considered separately. Suppose each has an acceleration of a m s^{-2}. Let the tension in the string be T N as shown in the diagram.

Using:

$$F = ma$$

Equation of motion for B is: $\quad 7 - T = 4a \quad (1)$

Equation of motion for A is: $\quad T = 3a \quad (2)$

Adding the simultaneous equations (1) and (2) gives:

$$7 = 4a + 3a$$
$$a = 1$$

The acceleration of either particle is 1 m s^{-2}.

Substituting $a = 1$ in equation (2) gives

$$T = 3 \times 1 = 3$$

The tension in the string is 3 N.

Example 11

A car of mass 1000 kg tows a caravan of mass 750 kg along a horizontal road. The engine of the car exerts a forward force of 2.5 kN. The resistances to the motion of the car and caravan are each k $\times$ their mass where k is constant. Given that the car accelerates at 1 m s^{-2} find the tension in the tow-bar.

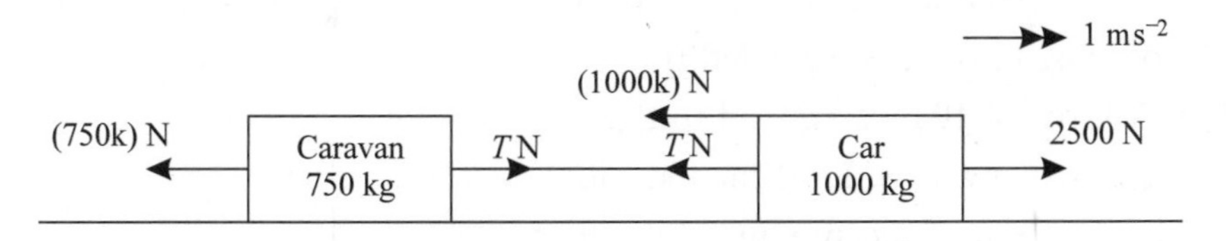

The diagram shows the information given.
Using $F = ma$ the equation of motion for the car is:

$$2500 - 1000\,\text{k} - T = 1000 \times 1$$
$$1000\,\text{k} + T = 2500 - 1000$$
$$1000\,\text{k} + T = 1500 \quad (1)$$

Equation of motion for the caravan is:

$$T - 750\,\text{k} = 750 \times 1$$

$$T - 750\,\text{k} = 750 \qquad (2)$$

Multiplying equation (1) by 3 gives:

$$3000\,\text{k} + 3T = 4500 \qquad (3)$$

Multiplying equation (2) by 4 gives:

$$4T - 3000\,\text{k} = 3000 \qquad (4)$$

Adding equations (3) and (4) gives:

$$7T = 7500$$

$$T = \frac{7500}{7} = 1071$$

The tension in the tow-bar is 1070 N.

Exercise 5C

Whenever a numerical value of g is required take $g = 9.8\,\text{m}\,\text{s}^{-2}$.

1 A car of mass 1000 kg is towing a caravan of mass 600 kg along a horizontal road. Given that the driving force produced by the engine is 400 N and that there is no resistance to motion find the tension in the tow-bar and the acceleration of the car.

2 A car of mass 850 kg is towing a caravan of mass 550 kg along a level road. The engine of the car exerts a forward force of 2 kN. The resistances to the motion of the car and caravan are proportional to their masses – in other words the resistances are k × their masses, where k is a constant. Given that the car accelerates at $0.5\,\text{m}\,\text{s}^{-2}$ find the tension in the tow-bar.

3 The diagram shows a particle A of mass 0.5 kg suspended by a vertical string. A particle B of mass 0.4 kg is suspended from A by means of another string. A force of 10 N is applied to the upper string and the particles move upwards. Find the tension in the lower string and the acceleration of the system.

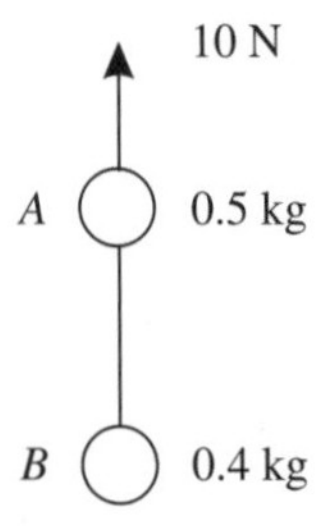

4 A car of mass 900 kg tows a caravan of mass 500 kg against resistances totalling 700 N. The resistances on the car and caravan are proportional to their masses – in other words the resistances are k × their masses where k is a constant. The car is accelerating at $0.8\,\text{m}\,\text{s}^{-2}$ along a horizontal road. By modelling the car and caravan as a pair of connected particles, find the driving force exerted by the engine and the tension in the tow-bar.

5 A car of mass 1000 kg exerts a driving force of 2.2 kN when pulling a caravan of mass 500 kg along a horizontal road. The car and caravan increase speed from rest to $4\,\mathrm{m\,s^{-1}}$ while travelling 16 m. Given that the resistances on the car and caravan are proportional to their masses, find these resistances and the tension in the tow-bar.

6 The diagram shows a block A of mass 100 kg suspended by a vertical cable. A block B of mass 150 kg is suspended from A by means of a second vertical cable. The blocks are raised 10 m in 10 seconds, starting from rest. Find the tension in each cable.

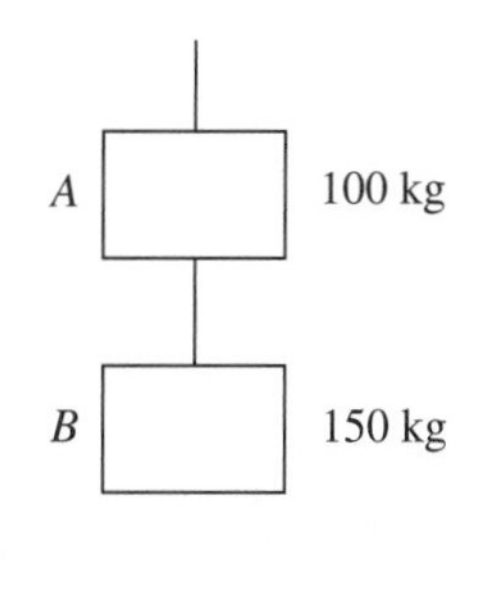

Problems involving pulleys

The figure shows two particles joined by a light inextensible string which passes over a fixed pulley. If the two particles are of different masses then the heavier particle will move vertically downwards and the lighter particle will move vertically upwards. However, the motions of the two particles are not independent because they are connected by the string. This is the motion of connected particles again although this time not along a single straight line.

Example 12

Particles of mass 5 kg and 2 kg are attached to the ends of a light inextensible string which passes over a smooth fixed pulley. The system is released from rest. Find the acceleration of the system and the distance moved by the 5 kg mass in the first 3 seconds of the motion. (Assume that neither particle reaches the pulley.)

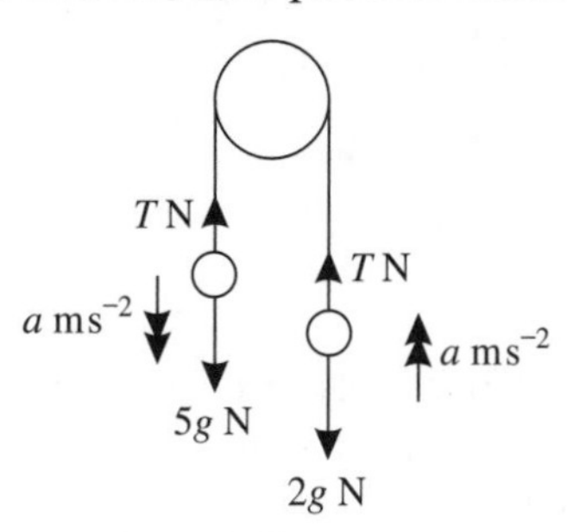

Since the particles are released from rest the heavier particle will move downwards. Let the particle's acceleration be $a\,\mathrm{m\,s^{-2}}$. As the particles are joined by a light inextensible string, the lighter mass will have an acceleration $a\,\mathrm{m\,s^{-2}}$ upwards.

Let the tension in the string be T N. It will be the same throughout the string as the pulley is smooth.

So for the motion of the particle of mass 5 kg use:

$$F = ma$$

Giving: $$5g - T = 5a \qquad (1)$$

And for the motion of the particle of mass 2 kg use:

$$F = ma$$

Giving: $$T - 2g = 2a \qquad (2)$$

Adding equations (1) and (2) gives:

$$5g - 2g = 5a + 2a$$

$$7a = 3g$$

$$a = \frac{3g}{7} = 3 \times \frac{9.8}{7} = 4.2$$

The acceleration is 4.2 $\mathrm{m\,s^{-2}}$.

To find the distance moved by the 5 kg particle in 3 seconds use the constant acceleration equations.

Known quantities are:

$$u = 0\,\mathrm{m\,s^{-1}}$$

$$a = 4.2\,\mathrm{m\,s^{-2}}$$

$$t = 3\,\mathrm{s}$$

Using: $$s = ut + \tfrac{1}{2}at^2$$

Gives: $$s = 0 \times 3 + \tfrac{1}{2} \times 4.2 \times 3^2$$

$$s = 18.9$$

The 5 kg mass moves 18.9 m in the first 3 seconds.

Example 13

Two particles P and Q of masses 6 kg and 3 kg respectively are connected by a light inextensible string. Particle P rests on a rough horizontal table. The string passes over a smooth pulley fixed at the edge of the table and Q hangs vertically. The coefficient of friction between P and the table is $\frac{1}{3}$. The system is released from rest. Find in terms of g (a) the acceleration of Q (b) the tension in the string and (c) the force exerted on the pulley.

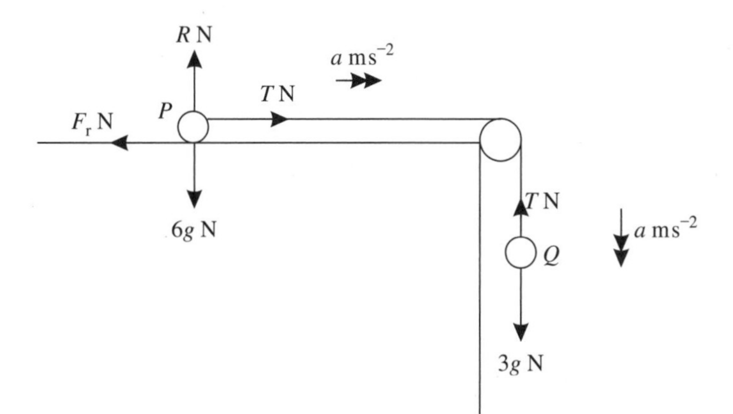

The tension in each part of the string is the same since the pulley is smooth. Let this tension be T N.

Let the acceleration of the particles be a m s^{-2} in the directions shown in the diagram.

(a) Particle P does not move vertically. So resolving vertically gives: $R - 6g = 0$.

As P is moving along the table, friction is limiting.

So: $F_r = \mu R = \frac{1}{3} \times 6g = 2g$

Using: $F = ma$ along the table

Gives: $T - F_r = 6a$

$$T - 2g = 6a$$

Resolving vertically for Q using: $F = ma$

Gives: $3g - T = 3a$ (2)

Adding equations (1) and (2) gives:

$$3g - 2g = 6a + 3a$$
$$9a = g$$
$$a = \frac{g}{9}$$

The acceleration of Q is $\frac{g}{9}$ m s^{-2} downwards.

(b) Substituting $a = \frac{g}{9}$ in equation (1) to find the tension in the string gives:

$$T - 2g = 6 \times \frac{g}{9}$$
$$T = 2g + \frac{2g}{3}$$
$$T = \frac{8g}{3}$$

The tension in the string is $\frac{8g}{3}$ N.

(c) To find the force exerted on the pulley consider the horizontal and vertical parts of the string separately. The tension in each part of the string is the same.

By Newton's third law, there will be equal but opposite tensions at the pulley ends of the strings:

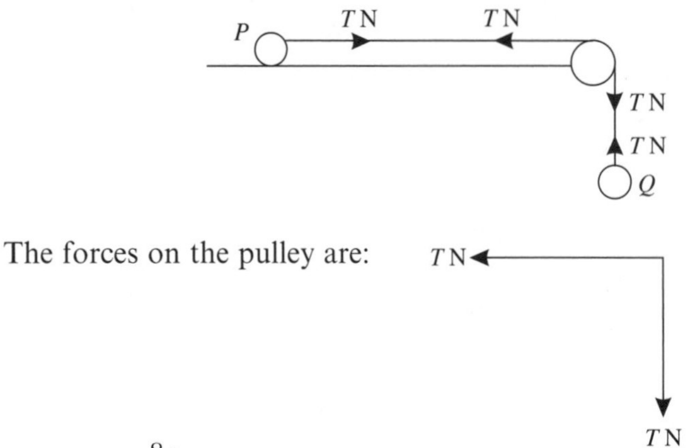

where $T = \dfrac{8g}{3}$ from (b) above.

To calculate the resultant apply the triangle law of addition (see chapter 2).

This gives:

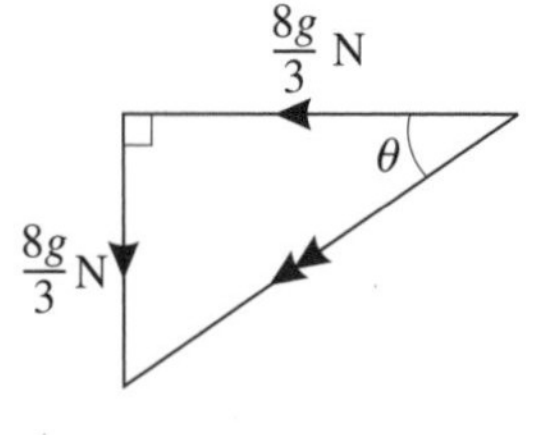

$$\text{Magnitude of resultant} = \sqrt{\left[\left(\frac{8g}{3}\right)^2 + \left(\frac{8g}{3}\right)^2\right]} = \frac{8g}{3}\sqrt{2}$$

$\theta = 45°$ as the triangle is isosceles.

The resultant force on the pulley is of magnitude $\dfrac{8g}{3}\sqrt{2}$ N in a direction 45° below the horizontal.

Example 14

A particle P of mass 5 kg lies on a smooth inclined plane of angle $\theta = \arcsin \frac{3}{5}$ (that is $\sin \theta = \frac{3}{5}$). Particle P is connected to a particle Q of mass 4 kg by a light inextensible string which lies along a line of greatest slope of the plane and passes over a smooth peg. The system is held at rest with Q hanging vertically 2 m above a horizontal plane. The system is now released from rest. Assuming P does not reach the peg, find, to 3 significant figures:

A **smooth peg** behaves in exactly the same way as a smooth pulley.

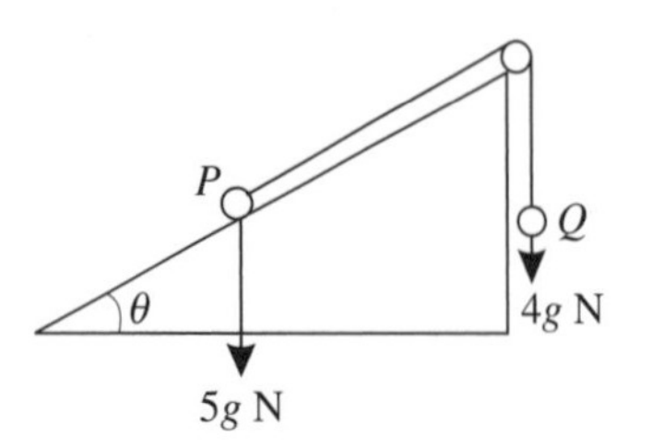

(a) the initial acceleration of Q
(b) how long it takes for Q to hit the horizontal plane
(c) the total distance that P moves up the plane.

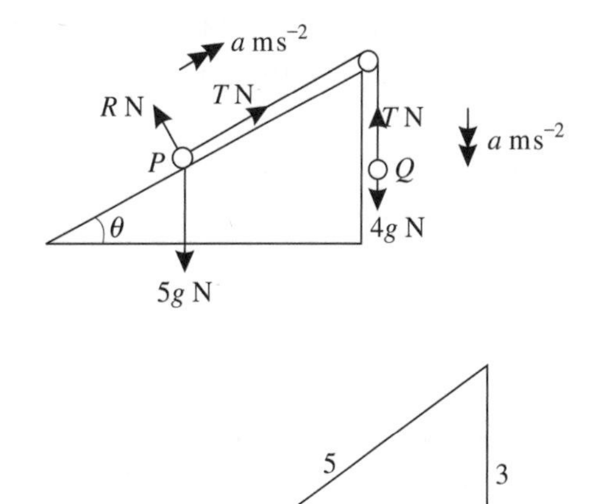

Using $\sin\theta = \frac{3}{5}$, from the (3, 4, 5) triangle, $\cos\theta = \frac{4}{5}$.

(a) As the peg is smooth, the tension is the same throughout the string. Let this tension be T N.

Let the acceleration of Q be a m s^{-2} downwards.

The acceleration of P is a m s^{-2} up the plane.

The plane is smooth so there is no friction force.

The weight of P has component $5g\sin\theta$ N parallel to the plane.

Using: $F = ma$

For P, parallel to the plane the equation of motion is:

$$T - 5g\sin\theta = 5a$$

$$T - 5g \times \frac{3}{5} = 5a$$

$$T - 3g = 5a \qquad (1)$$

For Q vertically the equation of motion is:

$$4g - T = 4a \qquad (2)$$

Adding equations (1) and (2) gives:

$$4g - 3g = 5a + 4a$$

$$a = \frac{g}{9} = \frac{9.8}{9} = 1.088$$

The initial acceleration is 1.09 m s^{-2}.

(b) Using: $s = ut + \frac{1}{2}at^2$

Known quantities are $s = 2$ m, $u = 0$, and $a = \frac{9.8}{9}$ m s^{-2}

Gives: $$2 = 0 \times t + \tfrac{1}{2} \times \frac{9.8}{9}t^2$$

$$t^2 = \frac{2 \times 2 \times 9}{9.8}$$

$$t = 1.916$$

Q takes 1.92 s to reach the horizontal plane.

(c) To calculate the total distance P travels up the plane before coming to rest separate P's motion up the plane into two parts, AB and BC.

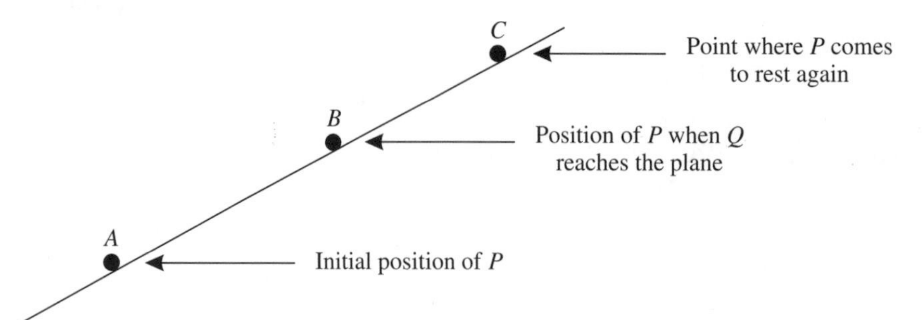

AB is the distance moved from the start of motion to the moment when Q reaches the plane.

BC is the distance moved from the moment Q reaches the plane until P comes to rest.

From A to B motion is as connected particles. Because Q descends 2 m before hitting the plane, P must travel the same distance so $AB = 2$ m.

Finding the speed of P when it reaches B will give the initial speed for the motion from B to C.

Using: $v^2 = u^2 + 2as$

Known quantities are $u = 0$, $a = 1.088\,\text{m s}^{-2}$ and $s = 2\,\text{m}$

Gives: $v^2 = 0 + 2 \times 1.088 \times 2$

$v = 2.086$

On reaching B, P has speed $2.086\,\text{m s}^{-1}$ up the plane and has travelled 2 m.

For the motion from B to C, a new diagram is needed because when Q reaches the plane the string becomes slack. The only forces on P are its weight and the reaction between P and the plane.

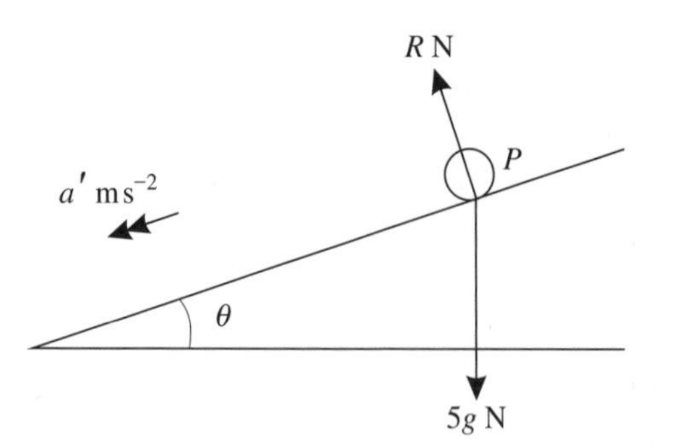

As the forces on P have changed, so will the acceleration.

Let the new acceleration be $a'\,\text{m s}^{-2}$.

Using: $F = ma$

The equation of motion parallel to the plane is:

$$5g \sin \theta = 5a'$$

$$a' = \frac{3}{5} g = 5.88 \text{ down the plane}$$

The motion from B to C is up the plane so the particle has an acceleration of -5.88 m s^{-2}.

Using: $v^2 = u^2 + 2as$ with $a = -5.88 \text{ m s}^{-2}$, $u = 2.086 \text{ m s}^{-1}$ $v = 0 \text{ m s}^{-1}$

Gives:

$$0 = 2.086^2 - 2 \times 5.88s$$

$$s = \frac{2.086^2}{2 \times 5.88} = 0.370$$

So the distance $BC = 0.37$ m.

Therefore the total distance moved by P is $AB + BC = 2.37$ m.

Exercise 5D

Whenever a numerical value of g is required take $g = 9.8 \text{ m s}^{-2}$.

1 Particles of mass 4 kg and 2 kg are attached to the ends of a light inextensible string which passes over a smooth fixed pulley. The system is released from rest. Find the acceleration of the system and the tension in the string.

2 Particles of mass 4 kg and 7 kg are attached to the ends of a light inextensible string which passes over a smooth fixed pulley. Find the acceleration of the system and the force exerted on the pulley.

3 Particles of mass $2M$ kg and M kg are connected by a light inextensible string which passes over a smooth fixed pulley. The system is released from rest with the hanging parts of the string vertical. Find the acceleration of the system in terms of g.

4 Two particles A and B are connected by a light inextensible string which passes over a smooth fixed pulley. A has mass 4 kg and descends 2.8 m in the first 2 seconds after the system is released from rest, the hanging parts of the string being vertical. Find the mass of B and the tension in the string.

5 Two particles P and Q of masses 3 kg and 6 kg respectively are attached to the ends of a light inextensible string. The string passes over a smooth fixed pulley. The system is released from rest with both masses a distance of 2 m above a horizontal floor. Find how long it takes for particle Q to hit the floor. Assuming particle P does not reach the pulley find its greatest height above the floor.

6 A rope is slung over a smooth beam. Two children are hanging on to the ends of the rope. The heavier child has a mass of 56 kg and descends with an acceleration of $2\,\mathrm{m\,s^{-2}}$. By modelling the children as a pair of connected particles find the tension in the rope and the mass of the lighter child.

7 Two particles P and Q of masses 7 kg and 2 kg respectively are connected by a light inextensible string. Particle P rests on a smooth table. The string passes over a smooth pulley fixed at the edge of the table and Q is hanging freely. The system is released from rest. Find the acceleration of P.

8 Given that the table for the system described in Question 7 is rough instead of smooth and that the coefficient of friction between P and the table is $\frac{1}{7}$, find the acceleration of the system and the tension in the string.

9 Two particles P and Q of masses 0.8 kg and 0.6 kg respectively are connected by a light inextensible string. Particle P rests on a rough horizontal table 0.7 m from a smooth fixed pulley at the edge of the table. The string passes over the pulley and Q hangs freely, 0.5 m above the floor. The system is released from rest. The coefficient of friction between P and the table is 0.25. Find (a) the acceleration of the system (b) the tension in the string (c) the speed with which Q hits the floor.

10 Two particles P and Q of masses 1 kg and 2 kg respectively are hanging vertically from the ends of a light inextensible string which passes over a smooth fixed pulley. The system is released from rest with both particles a distance of 1.5 m above a floor. When the masses have been moving for 0.5 s the string breaks. Find the further time that elapses before P hits the floor.

11 Two particles A and B of masses 0.5 kg and 0.7 kg respectively are connected by a light inextensible string. Particle A rests on a smooth horizontal table 1 m from a fixed smooth pulley at the edge of the table. The string passes over the pulley and B hangs freely 0.75 m above the floor. The system is released from rest with the string taut. After 0.5 s the string breaks. Find (a) the distance the particles have moved when the string breaks (b) the velocity of the particles when the string breaks (c) the further time that elapses before A reaches the floor, given that the table is 0.9 m high.

12

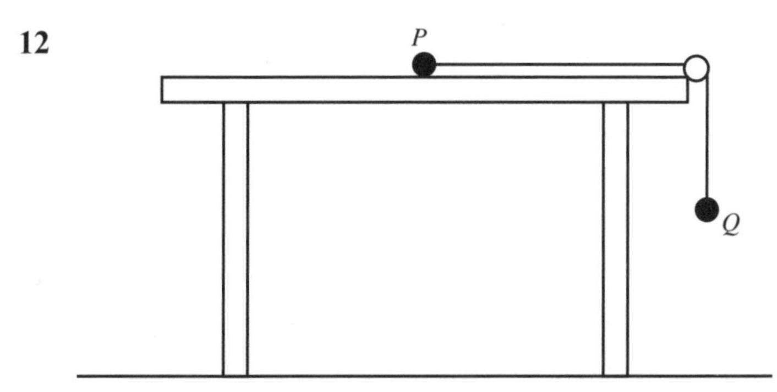

The diagram shows a particle P of mass 3 kg lying on a smooth horizontal table top which is 1.5 m above the floor. A light inextensible string of length 1 m connects P to a particle Q, also of mass 3 kg, which hangs freely over a small smooth pulley at the edge of the table. Initially P is held at rest at a point 0.5 m from the pulley. When the system is released from rest find

(a) the speed of P when it reaches the pulley,

(b) the tension in the string.

(c) In this problem several mathematical models have been used. Identify three of these and describe the assumptions which have been made in using these models.

13 A particle A of mass 2.5 kg which is at rest on a smooth inclined plane of angle 25° is connected to a particle B of mass 1.5 kg by a light inextensible string which lies along a line of greatest slope of the plane and passes over a fixed smooth pulley at the top of the plane. B hangs freely and the system is released from rest with the string taut. Find the acceleration of the system and the tension in the string. If the angle of inclination of the plane is increased to 45° find the new acceleration.

14 A particle A of mass 0.5 kg rests on a smooth inclined plane of angle arcsin $\frac{3}{5}$. A light inextensible string is attached to this particle and passes over a smooth pulley at the top of the plane. A particle B of mass m kg hangs freely from the other end of the string. The system is released from rest with the string taut. Particle B descends 1 m in 1 s. Find the value of m and the tension in the string.

15 A particle P of mass 5 kg rests on a rough inclined plane of angle arcsin $\frac{3}{5}$. The coefficient of friction between P and the plane is $\frac{1}{8}$. P is connected to a particle Q of mass 4 kg by a light inextensible string which lies along a line of greatest slope of the plane and passes over a smooth fixed pulley at the top of the plane. Q is hanging freely 2 m above a horizontal plane. The system is released from rest with the string taut. Assuming P does not reach the pulley find (a) the acceleration of the system (b) the time that elapses before Q hits the horizontal plane (c) the total distance P moves up the plane.

16 A particle A of mass 5 kg rests on a rough plane inclined at an angle of 30° to the horizontal. A string attached to A lies along a line of greatest slope of the plane and passes over a smooth pulley at the top of the plane. A particle B of mass 6 kg hangs vertically from the string 1 m above a horizontal plane. The system is released from rest with the string taut. If B takes 2 s to reach the horizontal plane, find the coefficient of friction between A and the inclined plane. Find also the total distance that A moves up the plane, assuming A does not reach the pulley.

5.3 Momentum and impulse

Suppose a particle of mass m moving in a straight line with constant acceleration increases its speed from u to v in time t. Then its acceleration can be found by using:

$$v = u + at$$

Or:

$$a = \frac{v - u}{t}$$

Applying the equation of motion to find the force required to produce this acceleration gives:

$$F = ma = m\frac{(v - u)}{t}$$

Or: $$Ft = mv - mu \qquad (1)$$

Momentum

For any moving particle, the quantity mass × velocity is called the **momentum** of the particle. Momentum is a vector quantity because it depends on the particle's velocity.

Impulse

If a constant force **F** is applied to a particle for a period of time t, then the quantity force × time ($\mathbf{F} \times t$) is called the **impulse**, **I**, of the force on the particle. Equation (1) can now be re-written as:

$$\mathbf{I} = m\mathbf{v} - m\mathbf{u}$$

where **u** is the initial velocity and **v** is the final velocity.

Or: Impulse = change in momentum.

For a mass in kg and a velocity in m s^{-1} force would be in newtons and hence impulse and momentum in newton-seconds (Ns).

Example 15

A particle of mass 5 kg is at rest on a smooth horizontal surface. A horizontal force of magnitude 4 N acts on the particle for 6 seconds. Find the final speed of the particle.

$\longrightarrow 0\ \text{ms}^{-1}$

Initial situation: 5 kg ○ $\longrightarrow$ 4 N

$\longrightarrow v\ \text{ms}^{-1}$

Final situation: 5 kg ○ $\longrightarrow$ 4 N

Let the final speed be $v\ \text{m s}^{-1}$. This will be in the same direction as the applied force.

Using: $$Ft = mv - mu$$

$$4 \times 6 = 5v - 0$$

$$v = 4.8$$

The final speed of the particle is $4.8\ \text{m s}^{-1}$.

In some situations, for example when a cricketer hits the ball, it is not possible to measure the length of time for which the bodies are in contact. Only the impulse of the force between the two bodies can be calculated.

Example 16

A ball of mass $\frac{1}{4}$kg hits a vertical wall with a horizontal speed of $30\,\text{m}\,\text{s}^{-1}$. It rebounds with a speed of $20\,\text{m}\,\text{s}^{-1}$. Find the impulse exerted by the wall on the ball.

In problems of this type it is helpful to draw diagrams to show the situation before and after the impact.

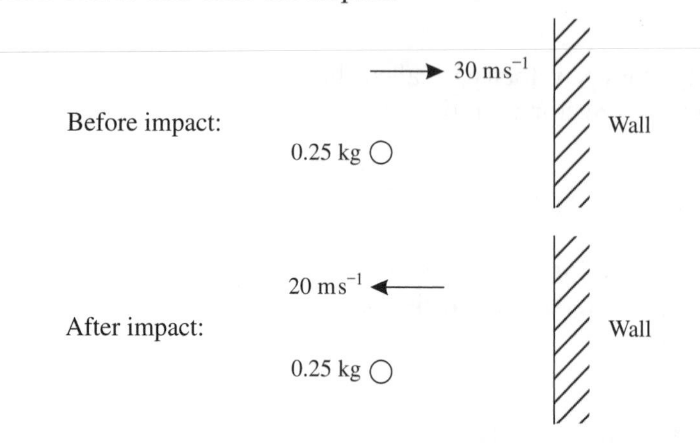

Because the velocity arrows in the two diagrams are in opposite directions, different signs must be used when substituting these velocities in the impulse-momentum equation.

Using: $I = mv - mu$

Taking the direction away from the wall to be positive gives:

$$I = 0.25 \times 20 - 0.25 \times (-30)$$
$$I = 12.5$$

The impulse exerted on the ball is 12.5 Ns.

By Newton's third law the force, and hence the impulse, on the ball is equal in magnitude to the force, or impulse, exerted on the wall. So the ball exerts an impulse of magnitude 12.5 Ns on the wall.

As momentum is a vector quantity, questions involving momentum may be expressed in vector form.

Example 17

At time $t = 0$ a particle of mass 8 kg is moving with a velocity of $(2\mathbf{i} + 2\mathbf{j})\,\text{m}\,\text{s}^{-1}$. Four seconds later its velocity is $(6\mathbf{i} - 4\mathbf{j})\,\text{m}\,\text{s}^{-1}$. Find the constant force acting on the particle.

Let the force be $(x\mathbf{i} + y\mathbf{j})$ N.

Using: $\mathbf{F}t = m\mathbf{v} - m\mathbf{u}$

Gives:

$$(x\mathbf{i} + y\mathbf{j}) \times 4 = 8(6\mathbf{i} - 4\mathbf{j}) - 8(2\mathbf{i} + 2\mathbf{j})$$
$$(x\mathbf{i} + y\mathbf{j}) = 2(6\mathbf{i} - 4\mathbf{j}) - 2(2\mathbf{i} + 2\mathbf{j})$$
$$x\mathbf{i} + y\mathbf{j} = 8\mathbf{i} - 12\mathbf{j}$$

The force is $(8\mathbf{i} - 12\mathbf{j})$ N.

Exercise 5E

1 Calculate the magnitude of the momentum of the following:

(a) a particle of mass 0.02 kg moving at $10\,\text{m}\,\text{s}^{-1}$

(b) a particle of mass 200 g moving at $100\,\text{m}\,\text{s}^{-1}$

(c) a car of mass 900 kg moving at $15\,\text{m}\,\text{s}^{-1}$

(d) a man of mass 80 kg running at $6\,\text{m}\,\text{s}^{-1}$

(e) a train of mass 150 tonnes moving at $20\,\text{m}\,\text{s}^{-1}$

(f) a car of mass 750 kg moving at $18\,\text{m}\,\text{s}^{-1}$.

2 A car of mass 800 kg decelerates from $20\,\text{m}\,\text{s}^{-1}$ to $5\,\text{m}\,\text{s}^{-1}$. Find the loss of momentum.

3 A car of mass 900 kg accelerates from rest to $18\,\text{m}\,\text{s}^{-1}$. Find the gain in momentum.

4 A particle of mass 0.2 kg is moving with speed $5\,\text{m}\,\text{s}^{-1}$. Its speed changes to $8\,\text{m}\,\text{s}^{-1}$ and its direction of travel is reversed. Find the change of momentum of the particle.

5 A toy car of mass 0.5 kg is pushed across a smooth horizontal floor by a horizontal force of 3 N for 4 s. Find the gain in momentum of the car. If the car starts from rest find its final speed.

6 A parcel of mass 10 kg is at rest on a smooth horizontal surface. It is pulled by a horizontal force of 15 N for 3 seconds. Model the car as a particle and hence find its final speed.

7 A particle of mass 6 kg is at rest on a smooth horizontal surface. It is acted on by a constant horizontal force for 5 s and then has a speed of $15\,\text{m}\,\text{s}^{-1}$. Find the magnitude of the force.

8 A particle of mass 10 kg is moving with a velocity of $7\mathbf{i}\,\text{m}\,\text{s}^{-1}$ when it is acted on by a force of $3\mathbf{i}$ N for 2 s. Find (a) the impulse given to the particle by the force (b) the final velocity of the particle.

9 A particle of mass 4 kg is moving with a velocity of $7\mathbf{j}\,\text{m}\,\text{s}^{-1}$; 3 s later it has a velocity of $-5\mathbf{j}\,\text{m}\,\text{s}^{-1}$. Find the constant force acting on the particle.

10 A particle of mass 0.2 kg is moving with a velocity of $8\mathbf{i}\,\text{m}\,\text{s}^{-1}$; t s later its velocity is $16\mathbf{i}\,\text{m}\,\text{s}^{-1}$. If the constant force acting on the particle has magnitude 0.4 N find the value of t.

11 A particle of mass 6 kg is moving with a velocity of $(3\mathbf{i} - 2\mathbf{j})\,\text{m s}^{-1}$; 3 s later it has a velocity of $(7\mathbf{i} + 3\mathbf{j})\,\text{m s}^{-1}$. Find the constant force acting on the particle.

12 A ball of mass 0.2 kg hits a vertical wall with a horizontal speed of $15\,\text{m s}^{-1}$. It rebounds with a speed of $12\,\text{m s}^{-1}$. Find the impulse exerted on the ball.

13 A smooth sphere of mass 1.5 kg hits a vertical wall with a horizontal speed of $2\,\text{m s}^{-1}$. It rebounds with a velocity of $1.5\,\text{m s}^{-1}$. Find (a) the impulse exerted on the sphere (b) the impulse exerted on the wall.

14 A ball of mass 0.5 kg is lying on the ground. It receives an impulse of 15 Ns from a bat. Model the ball as a particle and hence find the velocity with which the ball moves.

15 A ball of mass 0.4 kg is dropped from a height of 2.5 m. After hitting the ground it rises to a height of 1.8 m. By modelling the ball as a particle find (a) the speed with which the ball hits the ground (b) the speed with which the ball rebounds from the ground (c) the impulse the ball receives from the ground.

The principle of conservation of momentum

When two particles collide, by Newton's third law they exert equal and opposite forces, and hence impulses, on each other.

Consider two particles of masses m_1 and m_2 moving in the same direction in a straight line on a smooth horizontal surface with speeds u_1 and u_2 respectively where $u_1 > u_2$. The particles collide. Let their speeds after impact be v_1 and v_2 respectively in the same direction as u_1 and u_2 and let the impulse created by the impact be I.

Before impact: $\longrightarrow u_1$ $\longrightarrow u_2$

m_1 ○ m_2 ○

To show the impulse when the particles are in contact it is necessary to draw the particles a small distance apart.

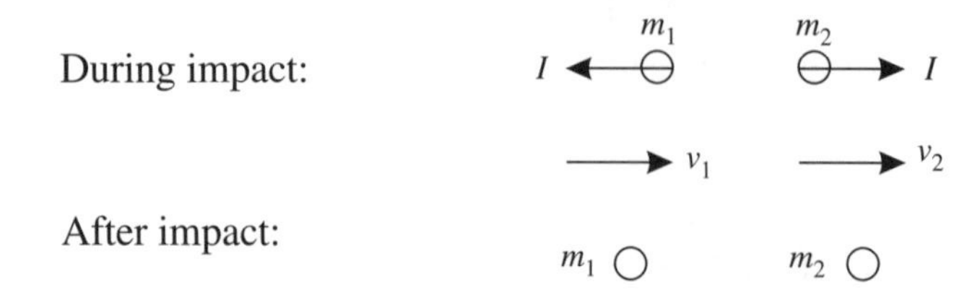

Because: Impulse = change of momentum

for each particle in turn use:

$$I = mv - mu$$

Take left $\rightarrow$ right to be the positive direction:

for m_1: $$-I = m_1v_1 - m_1u_1 \quad (1)$$

for m_2: $$I = m_2v_2 - m_2u_2 \quad (2)$$

Adding equations (1) and (2) gives:

$$0 = m_1v_1 - m_1u_1 + m_2v_2 - m_2u_2$$

Rearranging gives:

$$m_1u_1 + m_2u_2 = m_1v_1 + m_2v_2$$

Or in words:

- **Total momentum before impact = total momentum after impact**

This is known as the **principle of conservation of momentum**.

Example 18

Particle A of mass 2 kg is moving with speed $3\,\text{m s}^{-1}$ on a smooth horizontal surface. Particle B of mass 3 kg is at rest on the surface. A collides with B. After impact A continues to move in the same direction but with a speed of $1\,\text{m s}^{-1}$. Find the speed of B.

Draw diagrams to show the situations before and after impact. Let the speed of B after impact be $v\,\text{m s}^{-1}$.

Before impact: $\longrightarrow 3\,\text{ms}^{-1}$ A ○ 2 kg; $\longrightarrow 0\,\text{ms}^{-1}$ B ○ 3 kg

After impact: $\longrightarrow 1\,\text{ms}^{-1}$ A ○ 2 kg; $\longrightarrow v\,\text{ms}^{-1}$ B ○ 3 kg

Take left $\rightarrow$ right to be the positive direction.
By the principle of conservation of momentum:

$$m_1u_1 + m_2u_2 = m_1v_1 + m_2v_2$$

So:
$$2 \times 3 + 3 \times 0 = 2 \times 1 + 3v$$
$$3v = 4$$
$$v = \tfrac{4}{3} = 1\tfrac{1}{3}$$

The speed of B after impact is $1\frac{1}{3}\,\text{m s}^{-1}$.

Example 19

Two particles P and Q of masses 3 kg and 5 kg respectively are moving towards each other along the same straight line with speeds $4\,\text{m s}^{-1}$ and $2\,\text{m s}^{-1}$ respectively. After impact the direction of motion of P is reversed and its speed is $2\,\text{m s}^{-1}$. Find the speed of Q.

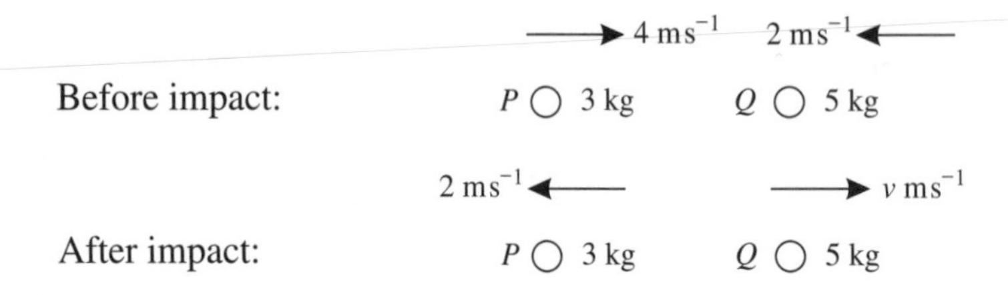

Let the speed of Q after impact be $v\,\text{m s}^{-1}$ as shown.

Take left → right to be the positive direction.

By the principle of conservation of momentum:

$$m_1u_1 + m_2u_2 = m_1v_1 + m_2v_2$$

So:

$$3 \times 4 - 5 \times 2 = -3 \times 2 + 5v$$

$$12 - 10 + 6 = 5v$$

$$v = \tfrac{8}{5} = 1\tfrac{3}{5}$$

The speed of Q after impact is $1\frac{3}{5}\,\text{m s}^{-1}$.

Example 20

A snooker ball P moving with speed $4\,\text{m s}^{-1}$ hits a stationary ball Q of equal mass. After the impact both balls move in the same direction along the same straight line, but the speed of Q is twice that of P. By modelling the balls as particles moving on a smooth horizontal surface find the speeds of the balls.

Let the balls have mass m kg and let the speed of P after impact be $v\,\text{m s}^{-1}$. The speed of Q will be $2v\,\text{m s}^{-1}$.

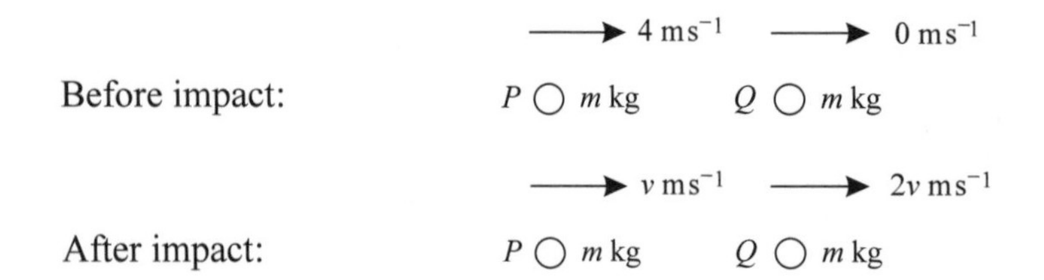

Take left → right to be the positive direction.

By the principle of conservation of momentum:

$$m_1u_1 + m_2u_2 = m_1v_1 + m_2v_2$$

So: $$4m + 0 = mv + 2mv$$

$$4 = 3v$$

$$v = \tfrac{4}{3} = 1\tfrac{1}{3}$$

After impact P has speed $1\frac{1}{3}\,\text{m s}^{-1}$ and Q has speed $2\frac{2}{3}\,\text{m s}^{-1}$.

Jerk in a string

The principle of conservation of momentum also applies to jerks in strings.

Consider two particles P and Q which are at rest on a smooth horizontal surface and are connected by a light inextensible string which is initially slack.

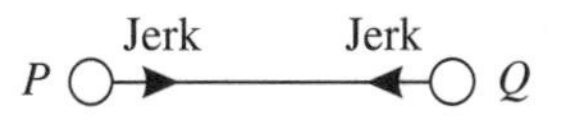

Suppose Q is given a velocity in the direction PQ. In time, the string will become taut. At the instant when the string becomes taut both particles will experience a jerk from the string.

By Newton's third law, the jerk experienced by P will be equal in magnitude but opposite in direction to the jerk experienced by Q. Therefore momentum is conserved.

Example 21

Two particles P and Q of mass 3 kg and 6 kg respectively are connected by a light inextensible string. Initially they are at rest on a smooth table with the string slack. Q is projected directly away from P with a speed of $3\,\text{m s}^{-1}$. Find their common speed when the string becomes taut.

Let the common speed of P and Q after the jerk be $v\,\text{m s}^{-1}$.

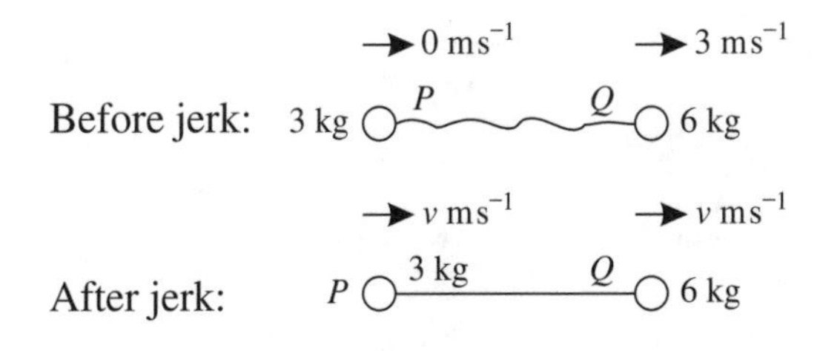

Take left → right to be the positive direction.

By the principle of conservation of momentum:

$$m_1u_1 + m_2u_2 = m_1v_1 + m_2v_2$$

So: $$3 \times 0 + 6 \times 3 = 3v + 6v$$

$$9v = 18$$

$$v = 2$$

The common speed of P and Q is $2\,\text{m s}^{-1}$.

Exercise 5F

The diagrams for questions 1 to 10 show the collision between two particles A and B. Find the unknown mass m or velocity v in each case.

	Before collision		After collision	
1	A: Mass 1 kg, → 3 ms^{-1}	B: Mass 0.5 kg, → 0 ms^{-1}	$A+B$: Mass 1.5 kg, → $v\text{ ms}^{-1}$	
2	A: Mass 2 kg, → 7.5 ms^{-1}	B: Mass 3 kg, → 0 ms^{-1}	$A+B$: Mass 5 kg, → $v\text{ ms}^{-1}$	
3	A: Mass 2 kg, → 7 ms^{-1}	B: Mass 3 kg, → 2 ms^{-1}	$A+B$: Mass 5 kg, → $v\text{ ms}^{-1}$	
4	A: Mass 2 kg, → 7 ms^{-1}	B: Mass 3 kg, ← 2 ms^{-1}	$A+B$: Mass 5 kg, → $v\text{ ms}^{-1}$	
5	A: Mass 100 g, → 5 ms^{-1}	B: Mass 200 g, → 2 ms^{-1}	A: Mass 100 g, → $v\text{ ms}^{-1}$	B: Mass 200 g, → 3 ms^{-1}
6	A: Mass 2 kg, → 5 ms^{-1}	B: Mass m g, → 3 ms^{-1}	A: Mass 2 kg, → 3 ms^{-1}	B: Mass m g, → 4 ms^{-1}
7	A: Mass 3 kg, → 2 ms^{-1}	B: Mass 2 kg, ← 6 ms^{-1}	A: Mass 3 kg, ← 4 ms^{-1}	B: Mass 2 kg, → $v\text{ ms}^{-1}$

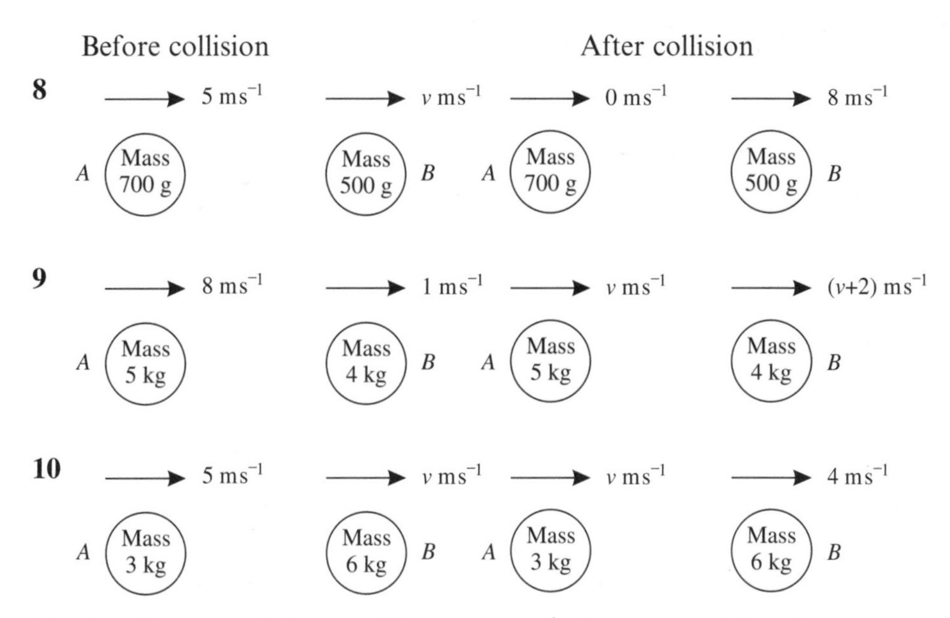

11 A railway truck is travelling at $6\,\text{m}\,\text{s}^{-1}$ along a straight piece of track when it hits an identical truck which is stationary. The two trucks couple together automatically. By modelling the trucks as particles find their common speed after the collision.

12 A small smooth sphere A of mass 2.5 kg lies at rest on a smooth horizontal table. A second small smooth sphere B of mass 1.5 kg is moving with speed $4\,\text{m}\,\text{s}^{-1}$ and collides directly with A. The two spheres coalesce – in other words after impact they move as a single body. Find their speed after impact.

13 Two particles P and Q of masses 3 kg and 2 kg respectively are at rest on a smooth horizontal surface. They are connected by a light inelastic string which is initially slack. P is projected away from Q with a speed of $10\,\text{m}\,\text{s}^{-1}$. Find the common speed of the particles after the string becomes taut and the impulse in the string when it jerks tight.

14 Particle A of mass 1 kg is at rest 0.2 m from the edge of a smooth horizontal 0.8 m high table. It is connected by a light inextensible string of length 0.7 m to a particle B of mass 0.5 kg. Particle B is initially at rest at the edge of the table closest to A but then falls off. Assuming B's initial horizontal velocity to be zero find the speed with which A begins to move.

15 A gun of mass 5 kg fires a bullet of mass 40 g. Given that the bullet leaves the gun with a speed of $500\,\text{m s}^{-1}$ find the initial speed of recoil of the gun.

16 A gun of mass 5000 kg discharges a shot of mass 25 kg. The gun recoils against a constant force of 10 kN which brings it to rest in 1.25 s. Find the speed of the shot.

17 A shell is travelling horizontally at $400\,\text{m s}^{-1}$ when it explodes into two pieces whose masses are in the ratio 2 : 3. The larger piece has a speed of $800\,\text{m s}^{-1}$ in the original direction of motion. Given that the smaller piece also moves horizontally after the explosion find its velocity.

18 A wooden peg of mass 0.4 kg is driven vertically into the ground by a mallet of mass 2 kg moving vertically downwards. The speed of the mallet just prior to impact is $8\,\text{m s}^{-1}$. After impact the mallet and peg remain in contact. By modelling the peg and mallet as particles find the speed with which the peg begins to enter the ground. If the ground offers a resistance to motion of 1 kN find how far the peg is driven into the ground due to the impact.

Exercise 5G Mixed questions

Whenever a numerical value of g is required take $g = 9.8\,m\,s^{-2}$.

1

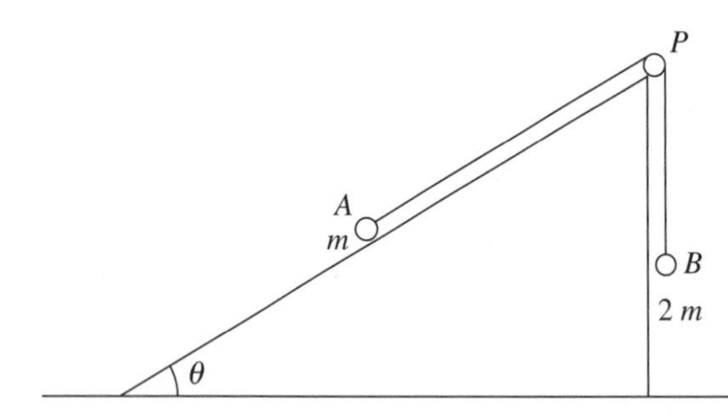

The diagram shows a particle A of mass m which can move on the rough surface of a plane inclined at an angle θ to horizontal ground, where $\theta = \arcsin 0.6$. A second particle B of mass $2m$ hangs freely attached to a light inextensible string which passes over a smooth pulley fixed at P. The other end of the string is attached to A. The coefficient of friction between A and the plane is $\frac{1}{4}$. B is initially hanging 1 m above the ground and A is 2 m from P. When the system is released from rest with the string taut A moves up the plane.

(a) Find the initial acceleration of A.

(b) Calculate the total distance moved by A before it first comes to rest.

2 A box of mass 5 kg is being pulled along rough horizontal ground by a rope inclined at 20° to the ground. When the tension in the rope is 10 N the box is moving at a constant speed.

(a) Calculate the coefficient of friction between the box and the ground.

The tension in the rope is now increased to 15 N.

(b) Calculate the acceleration of the box.

3 A particle P of mass $2m$ is moving in a straight line with constant speed $6u$. It collides with a particle Q of mass $4m$ moving in the same straight line towards P with constant speed $2u$. After the collision both particles move with speed v but in opposite directions.

(a) Find v in terms of u.

(b) Find, in terms of m and u, the magnitude of the impulse exerted by Q on P.

(c) State the direction of the impulse exerted by Q on P.

4

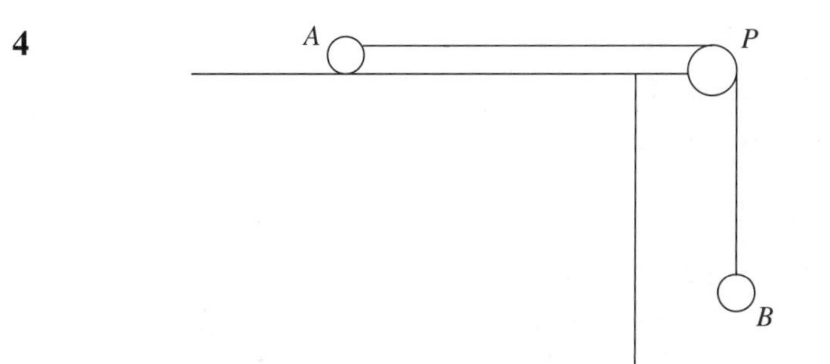

A small smooth pulley P is fixed at the end of a rough horizontal surface. A particle A of mass 0.6 kg rests on the surface at a distance of 2 m from P. A light inextensible string is attached to A and passes over the pulley. A particle B of mass 0.3 kg hangs freely from the other end of the string. When the system is released from rest with the string taut, particle B moves with an acceleration of magnitude $1\,\mathrm{m\,s^{-2}}$. Assuming that A has not reached P and B has not met any obstruction, calculate

(a) the tension in the string

(b) the coefficient of friction between the particle A and the surface

(c) the magnitude of the resultant force exerted by the string on the pulley.

T seconds after the system is released particle B has still not met any obstruction and particle A has just reached P.

(d) Calculate the value of T.

5 A small smooth sphere A of mass 0.1 kg is moving in a straight line with speed $4\,\mathrm{m\,s^{-1}}$ when it collides with another small smooth sphere B of mass 0.4 kg which is at rest. Immediately after the collision A and B are moving in opposite directions with speeds $x\,\mathrm{m\,s^{-1}}$ and $y\,\mathrm{m\,s^{-1}}$ respectively.

Given that $x : y = 2 : 3$,

(a) calculate the value of x

(b) calculate the magnitude of the impulse exerted on A by B.

6 A pebble of mass 0.5 kg is sliding in a straight line over rough horizontal ground. The coefficient of friction between the pebble and the ground is $\frac{1}{4}$. Calculate

(a) the frictional force acting on the pebble

(b) the retardation experienced by the pebble.

Given that the initial speed of the pebble is $6\,\mathrm{m\,s^{-1}}$,

(c) calculate the total distance travelled by the pebble before it comes to rest.

7 Two packets P and Q are connected by a rope which passes over a smooth fixed pulley. P has mass 4 kg and Q has mass M kg where $M > 4$. The system is released from rest with the string taut. The initial acceleration of each particle is $3\,\mathrm{m\,s^{-2}}$. By modelling the packets as particles and the rope as a light inextensible string, calculate

(a) the initial tension in the rope

(b) the value of M

(c) the initial force exerted by the rope on the pulley.

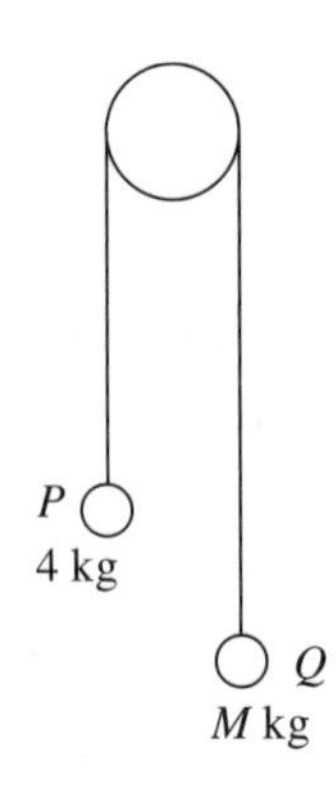

Given that during the first 2 s of the motion neither particle meets any obstruction,

(d) calculate the distance moved by P during the first 2 s.

(e) State how you have used the modelling assumption that the pulley is smooth.

8 A car of mass 800 kg tows a trailer of mass 300 kg along a straight road. The car and trailer are accelerating at a uniform rate of $2\,\text{m}\,\text{s}^{-2}$ and the total resistance acting on them is 550 N. In an initial model the car and trailer are modelled as particles, the tow-bar connecting them as a light inextensible horizontal rod, and the road is assumed to be horizontal.

(a) Calculate the driving force produced by the car's engine.

The resistance acting on the car is $800k$ newtons and the resistance acting on the trailer is $300k$ newtons.

(b) Calculate the value of the constant k.

(c) Calculate the tension in the tow-bar.

In a refined model the road is assumed to be inclined at an angle θ to the horizontal where $\sin\theta = 0.02$. The car and trailer are again modelled as particles and the tow-bar as a light inextensible rod parallel to the road. The resistance and driving force are unchanged.

(d) Calculate the new acceleration of the car and trailer.

9 A particle P is projected up a line of greatest slope of a rough plane inclined at an angle $\arctan\frac{5}{12}$ to the horizontal. The coefficient of friction between the particle and the plane is $\frac{1}{3}$. The particle is projected from point A with a speed of $30\,\text{m}\,\text{s}^{-1}$ and comes to an instantaneous rest at point B.

(a) Show that, whilst P is moving up the plane, its acceleration is of magnitude $\frac{9}{13}g$ and is directed down the plane.

(b) Find the distance AB, in metres to 3 significant figures.

(c) Find the time, in seconds to 3 significant figures, taken for P to move from A to B.

(d) Find the speed, in $\text{m}\,\text{s}^{-1}$, of the particle when it returns to A.

SUMMARY OF KEY POINTS

1 **Newton's Laws of Motion**

1 A particle will only accelerate if it is acted on by a resultant force.

2 The force F applied to a particle is proportional to the mass m of the particle and the acceleration produced.

$$\mathbf{F} = m\mathbf{a}$$

3 The forces between two bodies in contact are equal in magnitude but opposite in direction.

2 **Momentum and Impulse**

The momentum of a particle of mass m moving with velocity $\mathbf{v}$ is $m\mathbf{v}$.

If a constant force $\mathbf{F}$ acts on a particle for time t, then the impulse of the force is equal to $\mathbf{F}t$.

Impulse = change of momentum

For a force in newtons and time in seconds, impulse and momentum are measured in newton-seconds.

3 **Conservation of momentum**

Momentum is conserved for two particles experiencing a collision or jerk where there are no external forces involved.

$$m_1u_1 + m_2u_2 = m_1v_1 + m_2v_2$$

Moments

6

6.1 Moment of a force

To describe a force acting on a particle you need to know the magnitude and the direction of the force. The effect of the force on the particle is to produce a translation – that is the particle moves in a straight line without turning.

When a force is applied to a rigid body the point of application of the force must be given as well as its magnitude and direction. The force may cause the body to experience a **rotational** motion with or without a translational motion – that is the body may now turn as well as or instead of moving in a straight line.

Suppose you have two forces of unequal magnitude but want them to produce the same **turning effect** about a fixed point. To achieve this, the point of application of the smaller force must be at a greater distance from the fixed point than the point of application of the larger force.

A see-saw can be used to explain this.

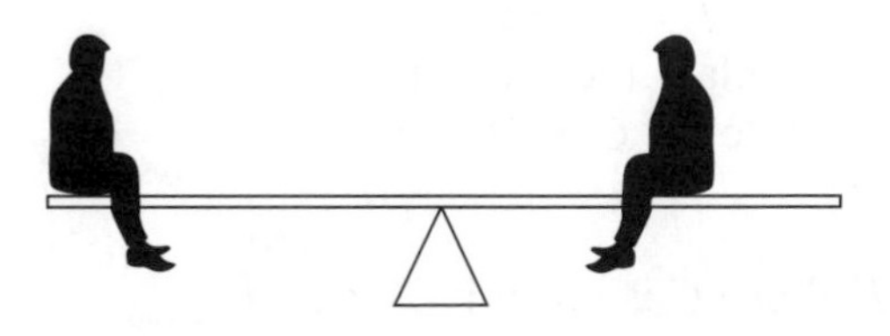

A see-saw is usually pivoted at its mid-point and can rotate about this pivot. If two children of unequal weight sit on the see-saw, the heavier child must sit nearer the pivot if the see-saw is to remain horizontal.

The **moment of a force** is a measure of its capability to turn the body on which it is acting.

The moment of the force F about a point P is the product of the magnitude of the force F and the perpendicular distance from the point P to the line of action of F.

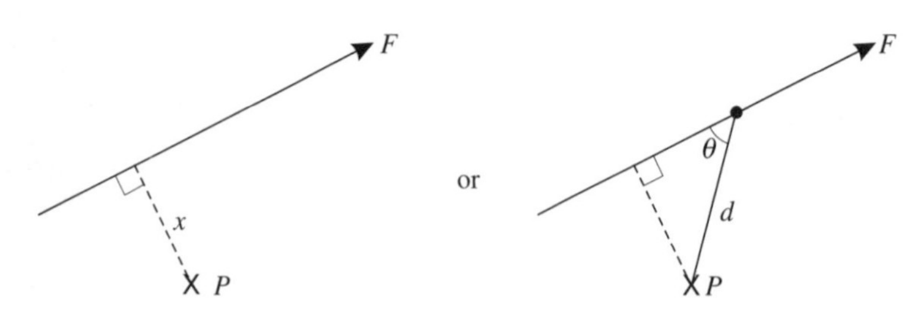

The moment of the force F about P is $F \times x = Fd \sin \theta$.

The moment of the force is measured in newton-metres (Nm) as the force is measured in newtons and the distance in metres.

Sense of rotation

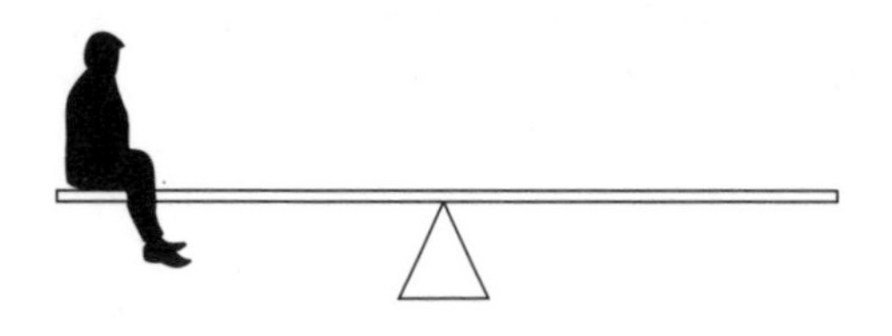

If one child sits on a see-saw on the left-hand side of the pivot, the weight of the child will turn the see-saw in the anticlockwise direction. If the child sits on the right-hand side, the see-saw will turn in a clockwise direction.

This clockwise or anticlockwise direction of the moment of a force is known as the **sense of rotation**. Use ↻ to show a clockwise direction and ↺ an anticlockwise direction.

Example 1

A light rod AB is 4 m long and can rotate in a vertical plane about a fixed point C where $AC = 1$ m. A vertical force F of magnitude 8 N acts in a direction perpendicular to AB. Calculate the moment of F about C when F is applied (a) at A (b) at B (c) at C as shown in the diagrams.

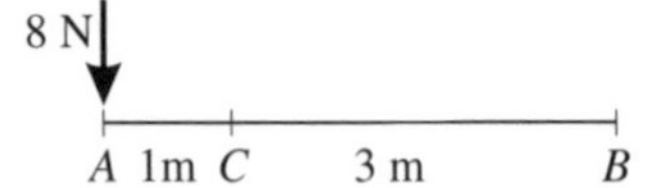

(a) When F is applied at A the moment of F about C is given by:

$$\text{Force} \times \text{perpendicular distance}$$

So:

$$\text{Moment} = 8 \times AC$$

Substituting in the value for AC gives:

$$\begin{aligned}\text{Moment} &= 8 \times 1\,\text{Nm} \\ &= 8\,\text{Nm anticlockwise}\end{aligned}$$

When F is applied at A, the moment is anticlockwise, because A is on the left-hand side of C and F is acting downwards.

(b) When F is applied at B:

The moment of F about C is given by:

$$\text{Force} \times \text{perpendicular distance}$$

So:

$$\text{Moment} = 8 \times CB$$

Substituting in the value for CB gives:

$$\text{Moment} = 8 \times 3\,\text{Nm}$$
$$= 24\,\text{Nm clockwise}$$

When F is applied at B, the moment is clockwise, because B is on the right-hand side of C and F is acting downwards.

(c) When F is applied at C:

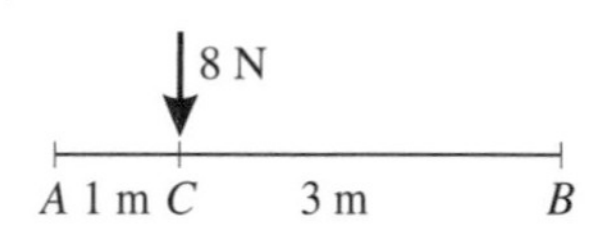

In this case, the perpendicular distance from C to the force is zero.

So the moment of F about C is:

$$8 \times 0\,\text{Nm} = 0\,\text{Nm}$$

When F is applied at C, there is no moment.

Sum of moments

When several forces, all lying in the same plane, act on a body the moments about any point can be added, as long as the sense of rotation for each moment is taken into account.

Example 2

Forces of magnitude 3 N, 4 N and 5 N are applied to a light rod AB of length 6 m as shown in the diagram. Calculate the sum of the moments of these forces about A.

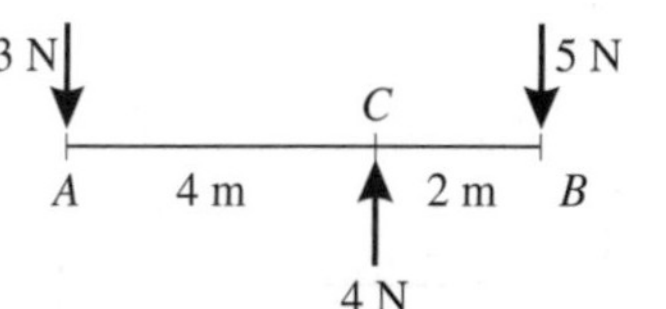

Moment of the 3 N force about A is:

$$3 \times 0\,\text{Nm}$$

Giving 0 Nm.

Moment of the 4 N force about A is:

$$4 \times 4\,\text{Nm}$$

Giving 16 Nm anti-clockwise ↺.
Moment of the 5N force about A is:

$$5 \times 6\,\text{Nm}$$

Giving 30 Nm clockwise ↻.

As the clockwise moment is larger than the anticlockwise moment, calculate the sum of the moments in the clockwise direction.

So the sum of the moments is:

$$(30 - 16)\,\text{Nm}$$

Giving 14 Nm clockwise ⤵.

The sum of the moments about A is 14 Nm clockwise.

Remember that the distance used to calculate a moment must be the *perpendicular* distance from the point to the force. If the length given is not perpendicular to the force, you need to know the angle between that length and the force to calculate the moment of the force.

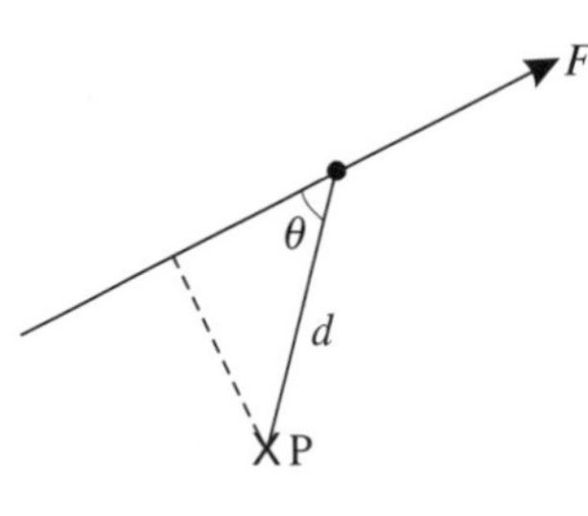

Use moment of the force $= Fd \sin\theta$.

Example 3

The following diagrams show forces acting on a lamina. In each case, calculate the sum of the moments of the forces about the point O.

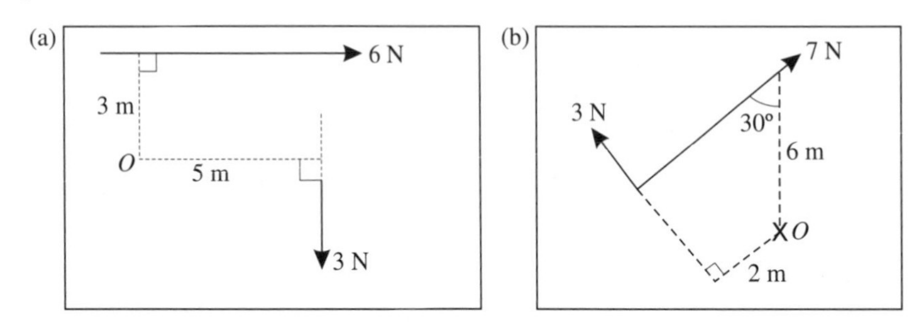

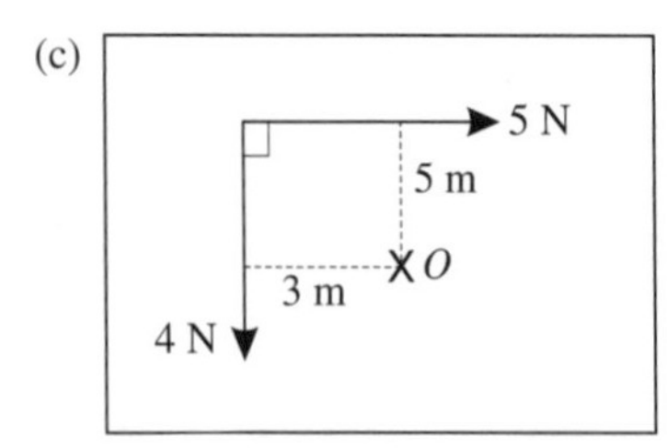

(a) As both forces have a clockwise moment, moment about O is:

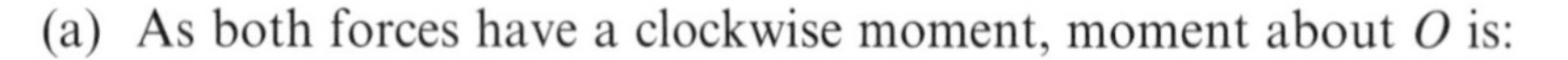

$$(6 \times 3 + 3 \times 5)⤵ = (18 + 15)⤵$$
$$= 33\,\text{Nm}⤵$$

The moment about O is 33 Nm, clockwise.

(b) Again both forces have a clockwise moment.

Moment of 7 N force about O is:

$$(7 \times 6 \sin 30) \circlearrowright = 21 \text{ Nm} \circlearrowright$$

So the total moment of the forces about O is:

$$(3 \times 2 + 21) \circlearrowright = 27 \text{ Nm} \circlearrowright$$

The moment about O is 27 Nm, clockwise.

(c) Here the 5 N force has a clockwise moment and the 4 N force has an anticlockwise moment.

So the moment about O is:

$$(5 \times 5 - 4 \times 3) \circlearrowright = 13 \text{ Nm} \circlearrowright$$

The moment about O is 13 Nm clockwise.

Example 4

(In this question **i** and **j** are unit vectors in the directions of the x-and y-axes respectively.)

The force $(6\mathbf{i} + 3\mathbf{j})$ N is applied at point A of a lamina, where A has position vector $(\mathbf{i} + 4\mathbf{j})$ m relative to a fixed origin O. Calculate the moment of the force

(a) about the origin O

(b) about the point B with position vector $(2\mathbf{i} - \mathbf{j})$ m.

First draw a clearly labelled sketch.

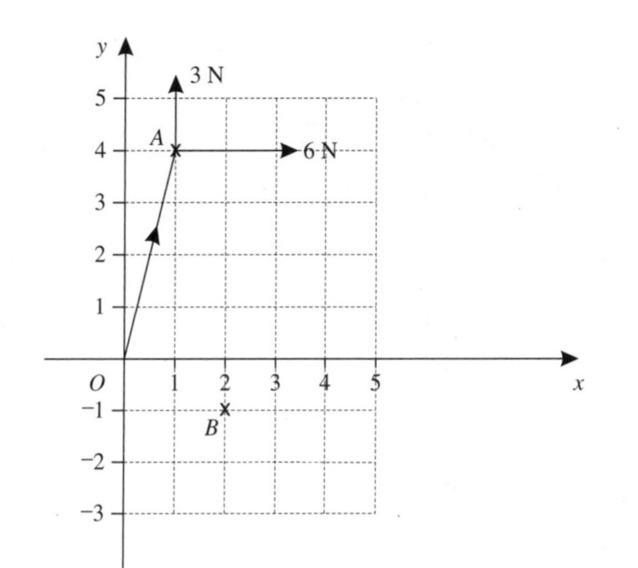

(a) It is convenient to use the components of the force which are parallel to the x-axis (6 N) and the y-axis (3 N), as shown in the diagram.
Use the diagram to read the following distances:

Perpendicular distance from O to 6 N force is 4 m
Perpendicular distance from O to 3 N force is 1 m
Moment of 6 N force is clockwise ↻
Moment of 3 N force is anticlockwise ↺

So the total moment about O is:

$$(6 \times 4 - 3 \times 1) \circlearrowright = 21 \text{ Nm} \circlearrowright.$$

The moment about O is 21 Nm clockwise.

(b) Use the diagram to calculate the following distances from B:

Perpendicular distance from B to 6 N force is 5 m
Perpendicular distance from B to 3 N force is 1 m
The moments of both forces are now clockwise

So the total moment about B is:

$$(6 \times 5 + 3 \times 1) \circlearrowright = 33 \text{ Nm} \circlearrowright$$

The moment about B is 33 Nm clockwise.

Exercise 6A

1 Calculate the moment about X of each of the following forces acting on a lamina. State the sense of each moment.

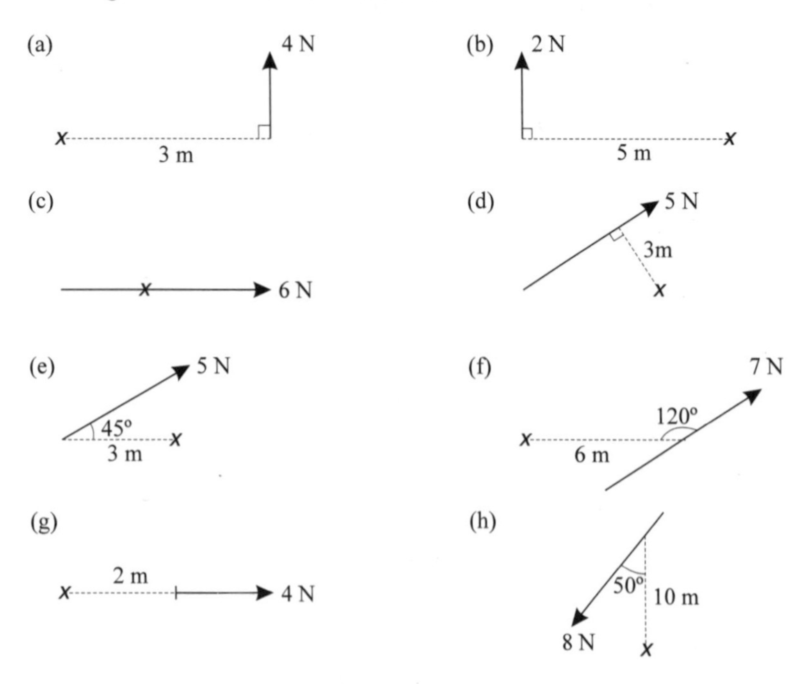

2 The following diagrams show sets of forces acting on a light rod. In each case calculate the sum of the moments about X stating the sense of each sum.

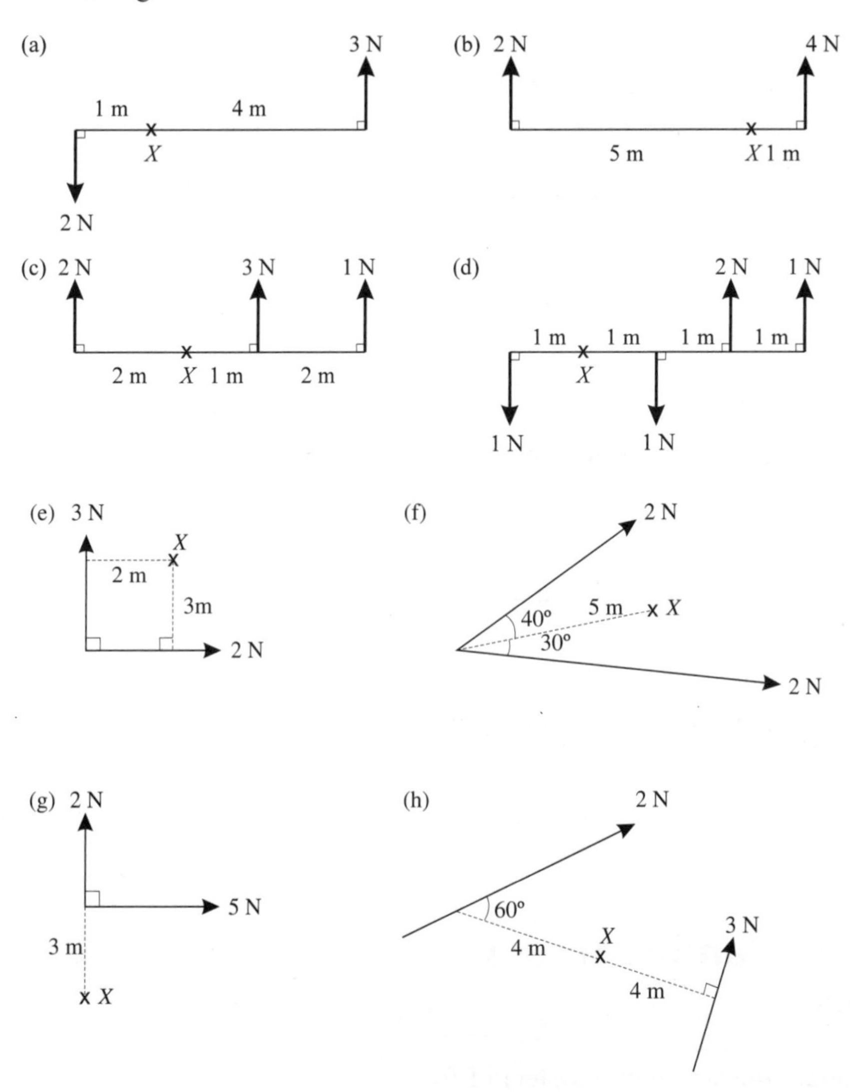

In questions 3 to 8, **i** and **j** are unit vectors in the directions of the x- and y-axes respectively.

3 The force $5\mathbf{j}$ N is applied at point A of a lamina where A has position vector $(2\mathbf{i} + 4\mathbf{j})$ m relative to a fixed origin. Calculate the moment of the force about the origin.

4 The force $(2\mathbf{i} + 4\mathbf{j})$ N acts at point P of a lamina where P has position vector $5\mathbf{j}$ m relative to a fixed origin. Calculate the moment of the force about the origin.

5 The force $2\mathbf{j}$ N acts at the point A of a lamina where A has position vector $6\mathbf{i}$ m relative to a fixed origin. Calculate the moment of the force about the point B which has position vector $4\mathbf{i}$ m.

6 The force $4\mathbf{i}$ N acts at the point A of a lamina where A has position vector $(2\mathbf{i} + 3\mathbf{j})$ m relative to a fixed origin. Calculate the moment of the force (a) about the origin (b) about the point with position vector $(4\mathbf{i} + 2\mathbf{j})$ m.

7 The force $(5\mathbf{i} + 3\mathbf{j})$ N acts at the point P of a lamina where P has position vector $(2\mathbf{i} - 2\mathbf{j})$ m relative to a fixed origin. Calculate the moment of the force (a) about the origin (b) about the point which has position vector $(-3\mathbf{i} + 2\mathbf{j})$ m.

8 The force $(3\mathbf{i} - 2\mathbf{j})$ N acts at point A of a lamina where A has position vector $(2\mathbf{i} + 3\mathbf{j})$ m relative to a fixed origin and the force $(4\mathbf{i} + 3\mathbf{j})$ N acts at the point B of the lamina where B has position vector $(3\mathbf{i} - 2\mathbf{j})$ m. Calculate the sum of the moments of these forces (a) about the origin (b) about the point with position vector $(4\mathbf{i} - 3\mathbf{j})$ m.

6.2 Equilibrium of a lamina under parallel forces

If a lamina is in equilibrium under the action of a system of forces the lamina will not turn about any point. This means the sum of the moments of the forces about any point must be zero. Account must be taken of the sense of the moments when calculating their sum. In chapter 4 it was shown that for a particle to be in equilibrium the resultant force must be zero.

This is also true for a lamina.

- **For a lamina in equilibrium under parallel forces:**
  (a) the component of the resultant force in any direction must be zero
  (b) the algebraic sum of the moments about any point must be zero.

It follows from (b) that the sum of the anticlockwise moments must equal the sum of the clockwise moments.

One of the forces acting on a lamina which is not light is its weight. The weight acts at the point called the **centre of mass** of the lamina. If the mass of the lamina is evenly distributed, the lamina is said to be **uniform**. The centre of mass of a uniform rod is at the mid-point of the rod.

Example 5

A uniform rod AB of length 4 m and mass 5 kg is pivoted at C where $AC = 1$ m. Calculate the mass of the particle which must be attached at A to maintain equilibrium with the rod horizontal.

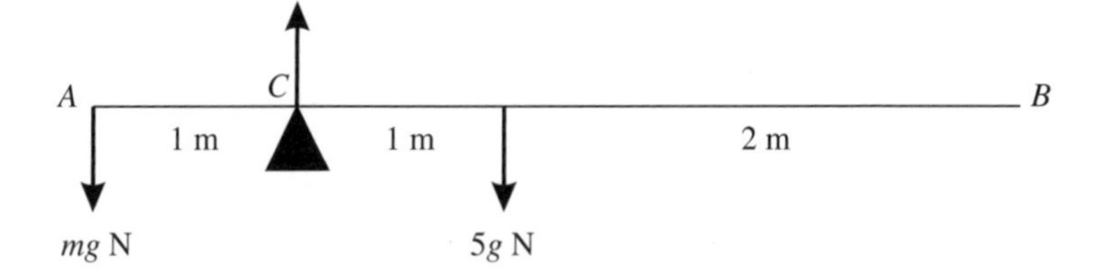

Let the mass of the particle at A be m kg.

The rod is uniform, so the weight of the rod acts at the mid-point of AB.

The pivot at C will exert a vertical force on the rod. The magnitude of this force is not known or required but the moment of this force about C is zero.

Taking moments about C gives:

$$\text{Clockwise moment about } C = 5g \times 1 \text{ Nm} \curvearrowright$$

$$\text{Anticlockwise moment about } C = mg \times 1 \text{ Nm} \curvearrowleft$$

For equilibrium:

$$\text{clockwise moment} = \text{anticlockwise moment}$$

So to calculate the mass of the particle required for equilibrium:

$$5g \times 1 = mg \times 1$$
$$m = 5$$

The required mass is 5 kg. This means that the mass to be attached at A must be 5 kg to keep the rod in equilibrium.

Example 6

A uniform beam AB of length 5 m and mass 30 kg rests horizontally on supports at C and D where $AC = BD = 1$ m.

A man of mass 75 kg stands on the beam at E where $AE = 2$ m. Calculate the magnitude of the reaction at each of the supports C and D.

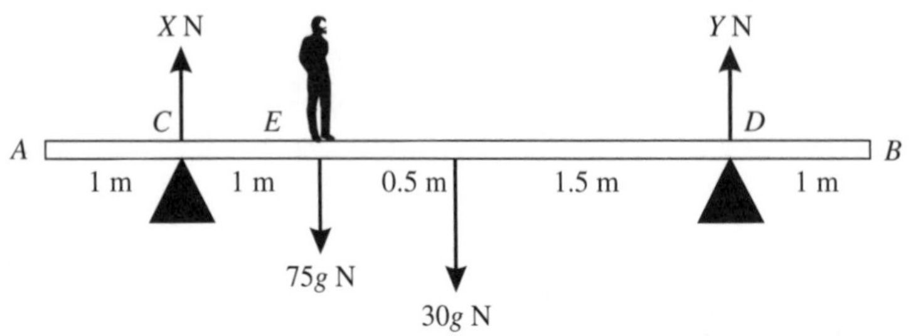

Let the reaction at C be X N and that at D be Y N.

To find the two unknown quantities X and Y two equations are needed:

(i) one obtained by taking moments about some convenient point
(ii) and the other obtained by resolving the forces vertically.

Taking moments about D and equating the clockwise and anticlockwise moments will give an equation which involves only one unknown quantity, X.

M(D): $$X \times 3 = 75g \times 2 + 30g \times 1.5$$
$$3X = 150g + 45g$$
$$X = \frac{195g}{3} = \frac{195 \times 9.8}{3}$$
$$= 637$$

The reaction at C has a magnitude of 637 N.

Resolving the forces vertically gives:

$$X + Y = 75g + 30g$$

Substituting for X gives:

$$Y = 105g - 637$$
$$= 392$$

The reaction at D has a magnitude of 392 N.

Alternatively, Y could be calculated by taking moments about C.

M(C): $$Y \times 3 = 75g \times 1 + 30g \times 1.5$$
$$3Y = 75g + 45g$$
$$Y = \frac{120g}{3} = 392$$

The reaction at D has a magnitude of 392 N.

Non-uniform rods

The centre of mass of a non-uniform rod is located at some point other than the mid-point of the rod. In some questions, such as example 7, the position of the centre of mass is given and you are required to find some other unknown quantity. In other questions, such as example 8, you will be given information to enable you to locate the centre of mass of the rod.

Example 7

Two children are sitting on a non-uniform plank AB of mass 24 kg and length 2 m. This plank is pivoted at M, the mid-point of AB. The centre of mass of AB is at C where AC is 0.8 m. Anne has mass 24 kg and sits at A. John has mass 30 kg.

Find where John must sit for the plank to be horizontal.

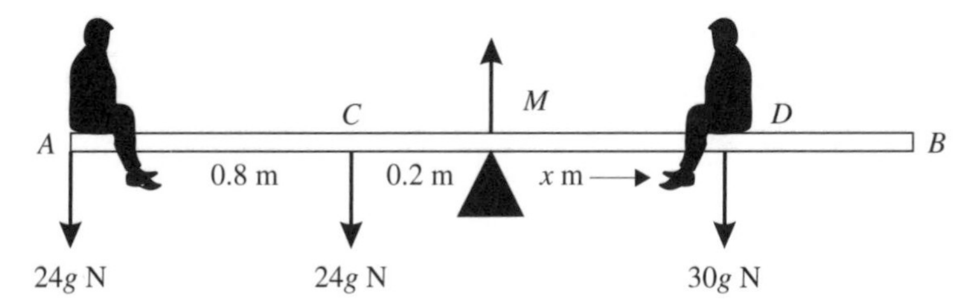

If you take moments about M, the unknown force at the pivot is not required.

M(M): $\qquad 24g \times 1 + 24g \times 0.2 = 30g \times x$

$$24 + 4.8 = 30x$$

$$x = \frac{28.8}{30} = 0.96$$

John should sit 0.96 m from M (or 0.04 m from B).

Example 8

A non-uniform rod AB of length 4 m and mass 5 kg is in equilibrium in a horizontal position resting on two supports at points C and D where $AC = 1$ m and $AD = 2$ m. The magnitude of the reaction at C is half the magnitude of the reaction at D. Find the distance of the centre of mass of the rod from A.

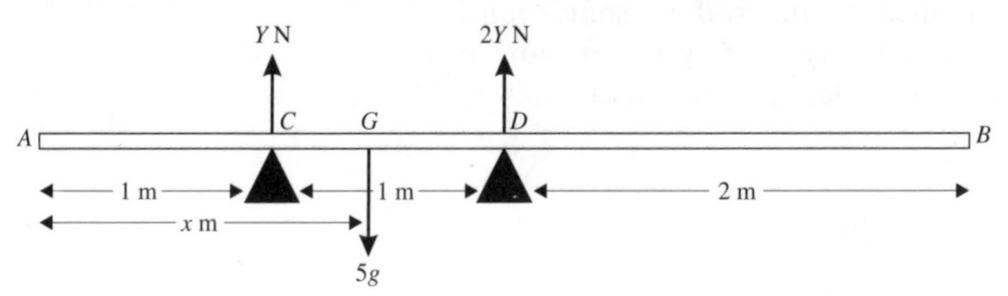

Let the centre of mass of the rod be at G where $AG = x$ m.

Let the reaction at C be Y N.

As the reaction at C is half the reaction at D, the reaction at D is $2Y$ N.

It is not possible to obtain a moments equation for the rod which does not include Y, so the value of Y must be found even though it is not demanded in the question.

Resolving the forces vertically gives:

$$Y + 2Y = 5g$$
$$Y = \tfrac{5}{3}g$$

Taking moments about A gives:

$$5g \times x = Y \times 1 + 2Y \times 2$$
$$5gx = 5Y$$
$$gx = Y$$

Substituting for Y gives:

$$gx = \tfrac{5}{3}g$$
$$x = \tfrac{5}{3}$$

The centre of mass of the rod is $1\tfrac{2}{3}$m from A.

Tilting

A heavy rod of mass M placed across two supports at A and B which are at the same horizontal level will be in equilibrium provided its centre of mass G is between A and B as shown below.

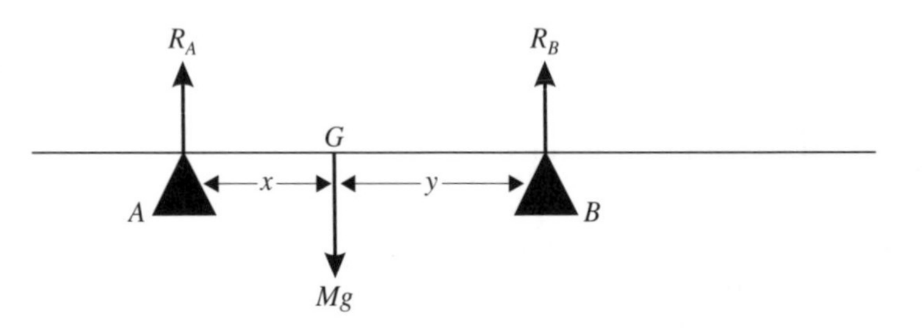

If an extra vertical force is applied to the rod at some point other than between A and B the rod may or may not remain in equilibrium, depending on the magnitude, F, of the force and the point of application.

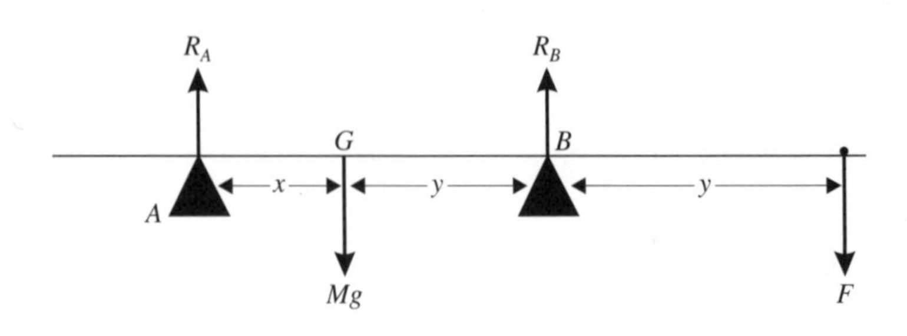

Without the extra force (that is, using the first diagram), taking moments about B gives:

$$R_A \times (x + y) = Mg \times y$$

With the extra force (that is, using the second diagram) and assuming that the rod is in equilibrium, taking moments about B gives:

$$R_A \times (x + y) + F \times z = Mg \times y$$

The right-hand sides of the two equations are identical, so we see that the reaction at A must have decreased as a result of applying the extra force. Also, the larger the magnitude of the extra force, the larger the decrease in the reaction at A will be. If the magnitude of the extra force is such that the rod is still in equilibrium but the reaction at A has decreased to zero, the rod is said to be on the point of *tilting* (or turning) about B.

Example 9

A uniform plank AB of mass 10 kg and length 2 m is in equilibrium in a horizontal position resting on two supports at points C and D where $AC = 0.5$ m and $AD = 1.3$ m. A boy of mass 30 kg stands on the plank at point E. The plank is on the point of tilting about D. By modelling the plank as a uniform rod and the boy as a particle, calculate the distance AE.

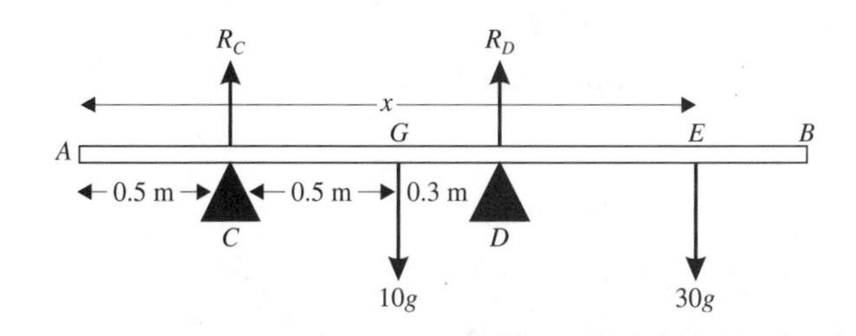

Let the distance AE be x m.

As the rod is on the point of tilting about D, $R_C = 0$.

Taking moments about D will give an equation which involves only one unknown quantity, x.

M(D):

$$30g \times (x - 1.3) = 10g \times 0.3$$

$$30g(x - 1.3) = 3g$$

$$(x - 1.3) = 0.1$$

$$x = 1.4$$

The distance AE is 1.4 m.

Exercise 6B

1 The following diagrams show a light rod in equilibrium in a horizontal position. In each case find the magnitudes of the forces X and Y and the distance dm as applicable.

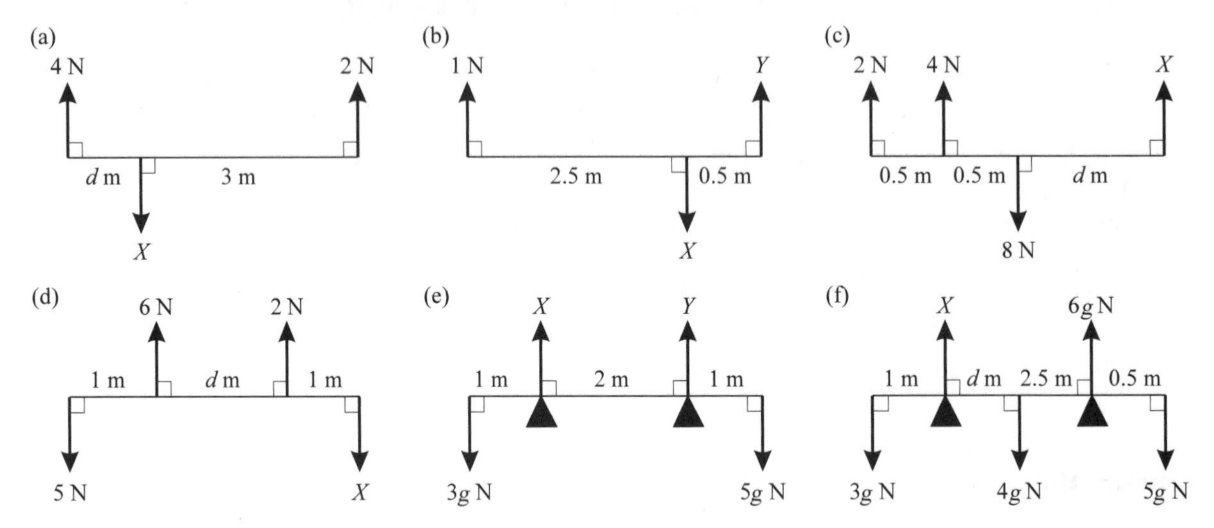

2 A uniform rod AB of length 5 m and mass 6 kg is pivoted at C where $AC = 1.5$ m. Calculate the mass of the particle which must be attached at A to maintain equilibrium with the rod horizontal.

3 A uniform beam AB of length 4 m and mass 21 kg has a particle of mass 15 kg attached at A and a particle of mass 18 kg attached at B. The beam is balanced horizontally on a single support at C. Find the distance AC.

4 A non-uniform rod AB of length 3 m and mass 4 kg rests horizontally on two supports at C and D where $AC = 0.5$ m and $BD = 1$ m. Given that the centre of mass of the rod is at M where $AM = 1.75$ m calculate the magnitudes of the reactions at the supports.

5 A rod AB of mass 6 kg and length 3 m has its centre of mass 1.2 m from A. It is suspended horizontally from the ceiling by two vertical strings one attached at A and the other at the point C where AC is 2.5 m. Find the magnitude of the tensions in the strings.

6 A non-uniform plank AB of length 5 m and mass 30 kg is pivoted at its mid-point. The plank is in equilibrium in a horizontal position when a child of mass 25 kg sits at A and a child of mass 35 kg sits at B. By modelling the plank as a rod and the children as particles find the distance of the centre of mass of the plank from A.

7 A uniform plank AB is 6 m long and is resting in a horizontal position on two supports at A and B. The mass of the plank is 20 kg. A body of mass 30 kg is placed on the plank 1 m from A and a second body of mass 30 kg is placed on the plank 4.5 m from A. Calculate the reaction at A and B.

8 A footbridge across a stream is constructed by placing a tree trunk AB of length 8 m and mass 90 kg on to supports at A and B so that the tree trunk is horizontal.

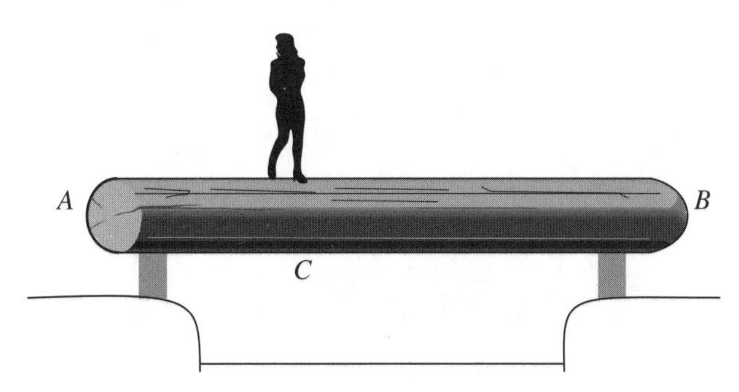

A woman of mass 60 kg stands on the trunk at C. The magnitude of the reaction at A is twice the magnitude of the reaction at B. By modelling the tree trunk as a uniform rod and the woman as a particle calculate (a) the magnitude of the reaction at A and the magnitude of the reaction at B (b) the distance AC. Describe the principal difference between your model and the real-life situation. How could your model be improved, using further information if necessary?

9 A uniform rod AB has length 6 m and mass 4 kg. It is resting in equilibrium in a horizontal position on supports at points X and Y where $AX = 2$ m and $AY = 4.5$ m. A particle of mass M kg is placed at point C where $AC = 5$ m. Given that the rod is on the point of tilting about Y, calculate the value of M.

10 A uniform rod AB of length 4 m and mass 2 kg is suspended in a horizontal position by two vertical strings attached at points P and Q where $AP = 1$ m and $AQ = 3$ m. When a particle of mass 3 kg is attached at point R of the rod, the rod is on the point of turning about P. Calculate the distance AR.

11 A non-uniform plank AB has length $5d$ and mass $8m$. It is in equilibrium in a horizontal position resting on supports at the points P and Q where $AP = 2d$ and $AQ = 4d$. A parcel of mass $6m$ is placed on the plank at B. The plank is on the point of tilting about Q. By modelling the plank as a rod and the parcel as a particle, calculate the distance of the centre of mass of the plank from A. Explain briefly the significance of modelling the parcel as a particle.

SUMMARY OF KEY POINTS

1 The moment of a force of magnitude F about a point P is given by:

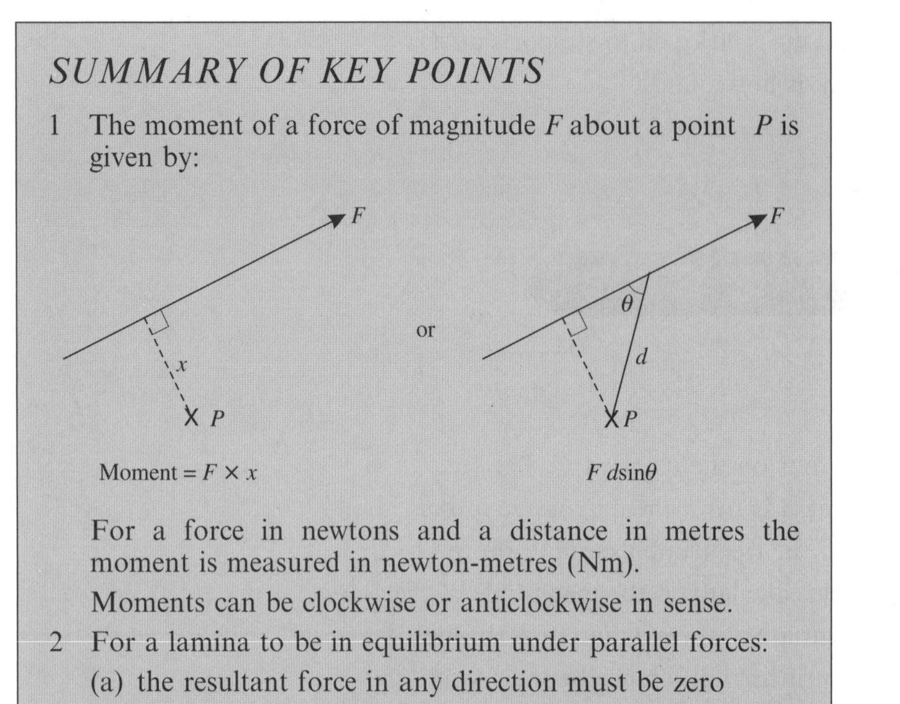

Moment $= F \times x$ $\quad$ or $\quad$ $F\,d\sin\theta$

For a force in newtons and a distance in metres the moment is measured in newton-metres (Nm).

Moments can be clockwise or anticlockwise in sense.

2 For a lamina to be in equilibrium under parallel forces:

(a) the resultant force in any direction must be zero

(b) the component of the resultant force in any direction must be zero

(c) the algebraic sum of the moments about any point must be zero.

3 A rod which is in equilibrium resting on two supports at the same horizontal level is said to be on the point of tilting about one support when the reaction at the other support has magnitude zero.

Review exercise

2

Whenever a numerical value of g is required take $g = 9.8$ m s^{-2}.

Note: $\dfrac{\sin\alpha}{\cos\alpha} = \tan\alpha$

1 A book of mass 2 kg rests on a rough plane inclined at an angle $\alpha°$ to the horizontal. Given that the coefficient of friction between the book and the plane is 0.2, and that the book is on the point of slipping down the plane, find, to the nearest degree, the value of α. [E]

2 A block of mass 3 kg rests on a rough, horizontal table. When a force of magnitude 10 N acts on the block at an angle of 60° to the horizontal in an upward direction, the block is on the point of slipping. Calculate, to 2 significant figures, the value of the coefficient of friction between the block and the table. [E]

3 A particle P of mass $7m$ is placed on a rough horizontal table, the coefficient of friction between P and the table being μ. A force of magnitude $2mg$, acting upwards at an acute angle α to the horizontal, is applied to P and equilibrium is on the point of being broken by the particle sliding on the table. Given that $\tan\alpha = \frac{5}{12}$, find the value of μ. [E]

4 (a) A book is placed on a desk lid which is slowly tilted. Given that the book begins to slide when the inclination of the lid to the horizontal is 30°, find the coefficient of friction between the book and the desk lid. [E]
(b) State an assumption you have made about the book when forming the mathematical model you used to solve part (a).

5 A particle is placed on a smooth plane inclined at 35° to the horizontal. The particle is kept in equilibrium by a horizontal force, of magnitude 8 N, acting in the vertical plane containing the line of greatest slope of the inclined plane through the particle. Calculate, in N to one decimal place,

(a) the weight of the particle

(b) the magnitude of the force exerted by the plane on the particle. [E]

6 A particle is suspended by two light inextensible strings and hangs in equilibrium. One string is inclined at 30° to the horizontal and the tension in that string is of magnitude 40 N. The second string is inclined at 60° to the horizontal. Calculate in N (a) the weight of the particle (b) the magnitude of the tension in the second string. [E]

7 A rough plane is inclined at an angle α to the horizontal, where $\tan \alpha = \frac{3}{4}$. A particle slides with acceleration 3.5 m s^{-2} down a line of greatest slope of this inclined plane. Calculate the coefficient of friction between the particle and the inclined plane. [E]

8 A particle starts from rest and slides with acceleration 3 m s^{-2} down a fixed plane which is inclined at 40° to the horizontal. Calculate, to 2 significant figures, the coefficient of friction between the particle and the plane. [E]

9 (a) Forces $(3\mathbf{i} + a\mathbf{j})$ newtons and $(b\mathbf{i} - 4\mathbf{j})$ newtons act on a body of mass 2 kg. The acceleration of the body is $(5\mathbf{i} - 3\mathbf{j}) \text{ m s}^{-2}$. Find the values of a and b. [E]

(b) Part (a) refers to 'the body'. What assumption did you make about 'the body' when solving the question?

10 A particle P of mass 2 kg is acted on by a force of $(10\mathbf{i} - 24\mathbf{j})$ N.

(a) Find, in m s^{-2}, the acceleration vector of P and the magnitude of this acceleration.

(b) Given that P starts from rest find, in m, the distance travelled by P in the first 3 seconds of the motion. [E]

11 A hot-air balloon, its occupants and ballast have total mass 2000 kg. The balloon is travelling horizontally with constant velocity. The occupants release 100 kg of ballast. Neglecting air resistance find, in $m\,s^{-2}$ to 2 significant figures, the magnitude of the acceleration of the balloon. [E]

12 A particle of mass 4 kg rests on a smooth horizontal plane. Forces of 8 N due east and 11 N due south are applied to the particle. Calculate the acceleration produced, giving its magnitude in $m\,s^{-2}$ to 2 significant figures and its direction as a bearing to the nearest degree. [E]

13 A block of mass 3 kg is pulled along a rough horizontal floor by a constant force of magnitude 20 N inclined at an angle of $60°$ to the upward vertical. The acceleration of the block has magnitude $2\,m\,s^{-2}$. Calculate, to 2 decimal places, the value of the coefficient of friction between the block and the floor. [E]

14 At a particular instant, the engine of a motor launch is producing a driving force of magnitude 6000 N and the launch is accelerating at $2\,m\,s^{-2}$. Given that the mass of the launch is 2300 kg find, in N, the magnitude of the force opposing the motion of the launch. [E]

15

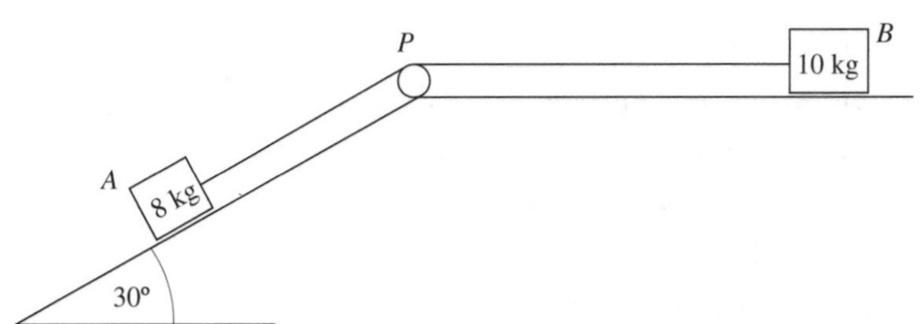

Two particles A and B, of mass 8 kg and 10 kg respectively, are connected by a light inextensible string which passes over a light smooth pulley P. Particle B rests on a smooth horizontal table and particle A rests on a smooth plane inclined at $30°$ to the horizontal with the string taut and perpendicular to the line of intersection of the table and the plane as shown above. The system is released from rest.

(a) Find the magnitude of the acceleration of particle B.

(b) Find, in newtons, the tension in the string.

(c) Find the distance covered by B in the first two seconds of motion, given B does not reach the pulley. [E]

16 (a) A tug, of mass 9000 kg, is pulling a boat, of mass 5000 kg, by means of an inextensible horizontal tow rope along a straight canal. The resistive forces opposing the motions of the tug and the boat are 1500 N and 800 N respectively. Calculate, in N, the tension in the rope when the tug is accelerating at $\frac{1}{2}$ m s^{-2}. [E]

(b) How did you model the tug and boat to answer part (a)?

(c) The tow rope is described as 'inextensible'. Do you think this is possible in real-life? Briefly justify your answer.

17 A particle A, of mass 0.8 kg, resting on a smooth horizontal table 1 m from the edge, is connected to a particle B, of mass 0.6 kg, which is 1 m from the ground, by a light inextensible string passing over a small smooth pulley fixed at the edge of the table. The system is released from rest with the horizontal part of the string perpendicular to the edge of the table, the hanging part vertical and the string taut. Calculate:

(a) the acceleration, in m s^{-2}, of A

(b) the tension, in N, in the string

(c) the speed, in m s^{-1}, of B when it hits the ground

(d) the time, in s, taken for B to reach the ground [E]

18 Two particles A and B, of masses 0.4 kg and 0.3 kg respectively, are connected by a light inextensible string. The particle A is placed near the bottom of a smooth plane inclined at 30° to the horizontal. The string passes over a small smooth light pulley which is fixed at the top of the inclined plane and B hangs freely. The system is released from rest, with each portion of the string taut and in the same vertical plane as a line of greatest slope of the inclined plane. Calculate:

(a) the common acceleration, in m s^{-2}, of the two particles

(b) the tension, in N, in the string.

Given that A has not reached the pulley, find:

(c) the time taken for B to fall 6.3 m from rest

(d) the speed that B has then acquired. [E]

19 Two particles A and B, of mass 6 kg and 1 kg respectively, are connected by a light inextensible string. Particle A is placed on a horizontal table. The string passes over a small smooth light pulley P fixed at the edge of the table and B hangs freely. The horizontal section of the string AP is perpendicular to the edge of the table and is of length 3 m. The particles are released from rest with both sections of the string taut and the section PB vertical.

If the table is smooth, find:

(a) the tension, in N, in the string

(b) the time, in seconds to 1 decimal place, taken by A to reach the pulley.

If instead, the table is rough and the coefficient of friction between A and the table is $\frac{1}{8}$, find:

(c) the acceleration, in m s^{-2}, of each particle

(d) the tension, in N, in the string. [E]

20 A car of mass 950 kg tows a caravan of mass 650 kg along a horizontal road. The engine of the car produces a tractive force of 2200 N.

(a) Find the acceleration of the car and the caravan when neither the car nor the caravan is subjected to any frictional resistance.

The car is now subjected to a constant frictional resistance of 150 N and the caravan to a constant frictional resistance of 200 N. The engine still produces a tractive force of 2200 N.

(b) Show, in two *separate* diagrams, *all* the horizontal forces acting on (i) the car (ii) the caravan.

(c) Find, in m s^{-2}, the magnitude of the acceleration of the car and the caravan.

(d) Find, to the nearest N, the magnitude of the tension in the coupling between the car and the caravan. [E]

21 Two particles A and B, of masses 0.4 kg and 0.3 kg respectively, are connected by a light inextensible string which passes over a smooth light fixed pulley. The particles are released from rest with the string taut and the hanging parts vertical. Calculate:

(a) the acceleration, in m s^{-2}, of A

(b) the tension, in N, in the string.

The particles continue to move in this system until the instant when they have each acquired speed 3.5 m s^{-1}. At this instant both A and B are 1.4 m above horizontal ground and the string is cut. Given that each particle now moves freely under gravity, find the difference in the times, measured from the instant when the string is cut, for A and B to reach the ground. [E]

22 A particle moves down a line of greatest slope of a rough plane which is inclined at 30° to the horizontal. The particle starts from rest and covers 3.5 m in time 2 s. Find the coefficient of friction between the particle and the plane. [E]

23 A particle *P*, of mass 0.3 kg, is moving in a straight line on a rough horizontal floor. The speed of *P* decreases from $7.5\,\text{m s}^{-1}$ to $4\,\text{m s}^{-1}$ in time *T* seconds. Given that the coefficient of friction between *P* and the floor is $\frac{1}{7}$ find:

(a) the magnitude of the frictional force opposing the motion of *P*

(b) the value of *T*. [E]

24 (a) A concrete slab, of mass 2 kg, falls from rest at a vertical height of 10 m above firm horizontal ground from which it does not rebound. Calculate the impulse of the force exerted on the ground by the slab and state the units in which your answer is measured. [E]

(b) State and justify briefly the assumptions you have made about the slab and the forces acting on it whilst it fell when answering part (a).

25 A railway truck, of mass 1500 kg and travelling with speed $6\,\text{m s}^{-1}$ along a horizontal track, collides with a stationary truck of mass 2000 kg. After the collision the two trucks move on together, coming to rest after 12 seconds. Calculate, in N, the constant force resisting their motion after the collision. [E]

26 (a) A ball of mass 0.3 kg is released from a point at a height of 10 m above horizontal ground. After hitting the ground the ball rebounds to a height of 2.5 m. Calculate, in N s, the magnitude of the impulse of the force exerted on the ball by the ground during the impact. [E]

(b) State two assumptions you have made about the ball and the forces acting on it in order to solve part (a).

27 A shell of mass 50 kg is fired with speed $560\,\text{m s}^{-1}$. Given that the shell is in the barrel of the gun for $\frac{1}{20}$ second, calculate the average force, in kN, exerted on the shell by the explosive charge. [E]

28 A pile driver of mass 350 kg drives vertically a pile of mass 630 kg into horizontal ground. The pile driver strikes the pile directly with speed $7\,\text{m s}^{-1}$ and does not rebound, so that the pile driver and pile move as one body into the ground.
(a) Show that the common speed of the pile driver and pile at the instant immediately after impact is $2.5\,\text{m s}^{-1}$.
(b) Calculate, in Ns, the impulse of the force exerted by the pile driver on the pile at impact.
The pile driver and pile penetrate 0.5 m into the ground. Assuming that the force exerted by the ground on the pile and pile driver during the motion is constant,
(c) calculate, in N to 2 significant figures, the value of this constant force. [E]

29 Two particles, A and B, of mass 0.2 kg and 0.5 kg respectively, are connected by a light inextensible string passing over a fixed smooth light pulley. The particles are released from rest with the string taut and the hanging parts vertical.
(a) Prove that the acceleration of B is of magnitude $4.2\,\text{m s}^{-2}$.
(b) Calculate, in N, the tension in the string.
At the instant when A and B are moving with speed $6.3\,\text{m s}^{-1}$, B strikes an inelastic horizontal floor from which it does not rebound. Calculate
(c) the magnitude of the impulse exerted on the floor by B, stating the units in which your answer is measured,
(d) the total time from the instant when the particles were first released until the instant when A first comes to instantaneous rest. (You may assume that A does not reach the pulley.) [E]

30 Two particles P and Q, of mass 2 kg and 3 kg respectively, are placed at rest at the points A and B respectively on a smooth horizontal table. An impulse of 8 Ns, in the direction $\overrightarrow{AB}$, is given to the particle P.
(a) Calculate the speed with which particle P begins to move.
Particle P collides with particle Q. Particle P moves off with speed $v\,\text{m s}^{-1}$ and particle Q moves off with speed $2v\,\text{m s}^{-1}$, both in the direction $\overrightarrow{AB}$.
(b) Calculate the value of v. [E]

31 A particle X, of mass 1 kg, moves with speed $3\,\text{m}\,\text{s}^{-1}$ along a smooth straight horizontal groove. It strikes a particle Y, of mass 2 kg, which is at rest in the groove. Immediately after the collision Y moves with speed $2u\,\text{m}\,\text{s}^{-1}$ and X moves with speed $u\,\text{m}\,\text{s}^{-1}$ in the *opposite* direction.

(a) Calculate the value of u.

(b) Calculate, in Ns, the magnitude of the impulse received by Y at the impact. [E]

32 A particle P, of mass 0.3 kg, moves in a straight line with speed $2\,\text{m}\,\text{s}^{-1}$ on a smooth horizontal plane. It collides with a stationary particle Q, of mass 0.2 kg. The two particles coalesce and move on together. Find the speed, in $\text{m}\,\text{s}^{-1}$, of the combined particle after the collision. [E]

33 A straight uniform rigid beam AB, of length 6 m and mass 15 kg, is supported at A and B and rests horizontally. A load of mass 60 kg is placed on the beam at the point C. Given that the magnitude of the force exerted on the support at A is twice the magnitude of the force exerted on the support at B, find the distance AC. [E]

34 (a) A uniform horizontal plank AB, of mass 30 kg and length 3 m, rests in equilibrium on two supports at C and D, where $AC = DB = 0.5$ m. A man of mass 76 kg stands on the plank at E, where $EB = 1$ m.
Find the magnitudes, in N, of the forces exerted by the plank on the supports at C and D. [E]

(b) How did you model the man and the plank in order to solve part (a)?

35 A footbridge across a stream consists of a uniform horizontal plank AB, of length 5 m and mass 140 kg, supported at the ends A and B. A man of mass 100 kg is standing at a point C on the footbridge. Given that the magnitude of the force exerted by the support at A is twice the magnitude of the force exerted by the support at B, calculate

(a) the magnitude, in N, of the force exerted by the support at B

(b) the distance AC. [E]

36 A force **R** acts on a particle, where $\mathbf{R} = (7\mathbf{i} + 16\mathbf{j})$ N.
Calculate

(a) the magnitude of **R**, giving your answer to 1 decimal place,

(b) the angle between the line of action of **R** and **i**, giving your answer to the nearest degree.

The force **R** is the resultant of two forces **P** and **Q**. The line of action of **P** is parallel to the vector $(\mathbf{i} + 4\mathbf{j})$ and the line of action of **Q** is parallel to the vector $(\mathbf{i} + \mathbf{j})$.

(c) Determine the forces **P** and **Q** expressing each in terms of **i** and **j**. [E]

37 Two horizontal forces, **P** and **Q**, act on a particle.
The force **P** is of magnitude 8 N and acts in the direction whose bearing is 330°.
The force **Q** is of magnitude 15 N and acts in the direction whose bearing is 060°.
Calculate the magnitude and the direction of the resultant of **P** and **Q**, giving the direction as a bearing to the nearest degree. [E]

38

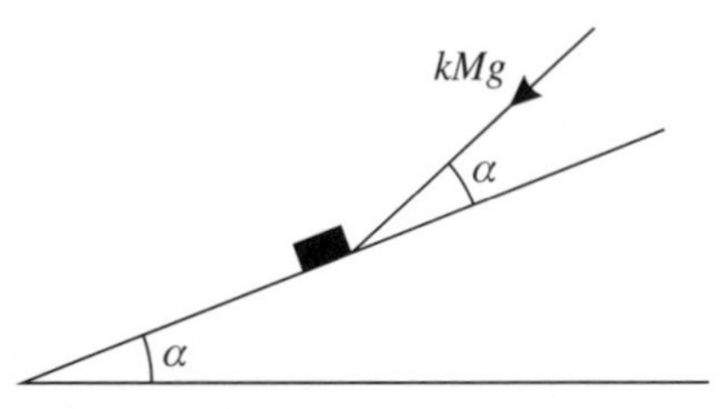

A rough slope is inclined at an angle α to the horizontal, where $\alpha < 45°$. A small parcel of mass M is at rest on the slope, and the coefficient of friction between the parcel and the slope is μ. A force of magnitude kMg, where k is a constant, is applied to the parcel in a direction making an angle α with a line of greatest slope, as shown in the diagram. The line of action of the force is in the same vertical plane as the line of greatest slope.
Given that the parcel is on the point of moving down the slope, show that

$$k = \frac{\mu \cos\alpha - \sin\alpha}{\cos\alpha - \mu \sin\alpha}$$

[E]

39 A particle of mass 0.3 kg lies on a smooth plane inclined at an angle α to the horizontal, where $\tan \alpha = \frac{3}{4}$. The particle is held in equilibrium by a horizontal force of magnitude Q newtons. The line of action of this force is in the same vertical plane as a line of greatest slope of the inclined plane. Calculate the value of Q, to one decimal place. [E]

40

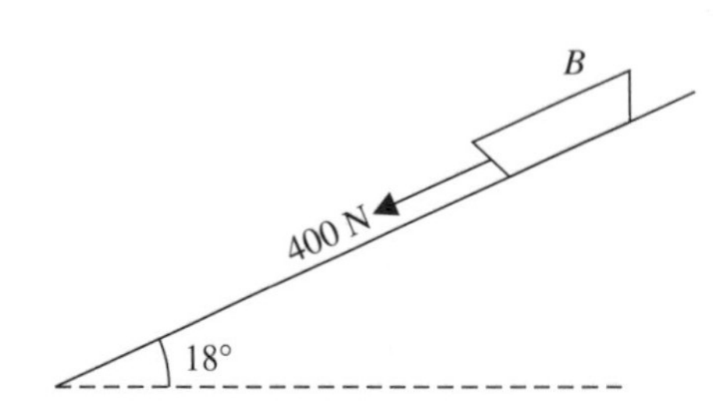

A small boat B of mass 100 kg is standing on a ramp which is inclined at 18° to the horizontal. A force of magnitude 400 N is applied to B and acts down the ramp as shown in the diagram. The boat is in limiting equilibrium on the point of sliding down the ramp. Find the coefficient of friction between B and the ramp, giving your answer to 2 decimal places. [E]

41

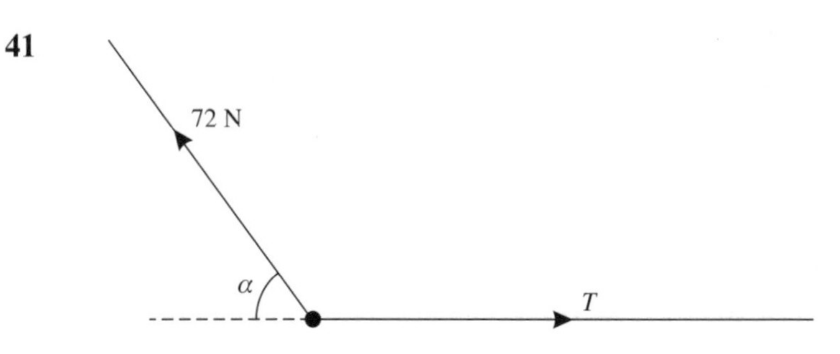

A body of mass 5 kg is held in equilibrium under gravity by two inextensible light ropes. One rope is horizontal, the other is at an angle α to the horizontal, as shown in the diagram. The tension in the rope inclined at α to the horizontal is 72 N. Find

(a) the angle α, giving your answer to the nearest degree,

(b) the tension T in the horizontal rope, giving your answer to the nearest N. [E]

42

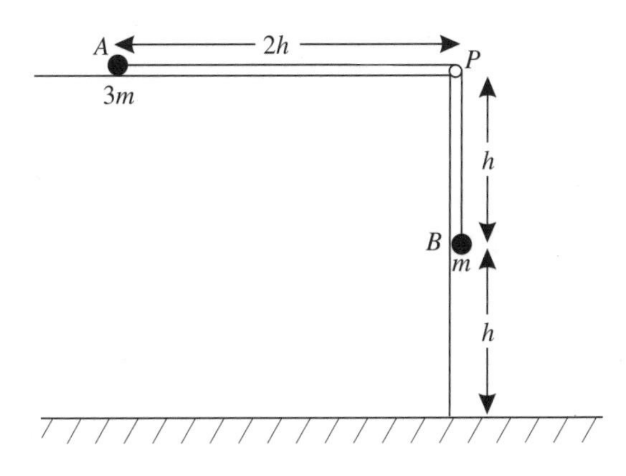

Particles A and B, of mass $3m$ and m respectively, are attached to the ends of a light inextensible string of length $3h$. The string passes over a small, smooth pulley P, fixed at the edge of a rough, horizontal table. The coefficient of friction between A and the table is μ.

The system is held at rest with the string taut, A being on the table and B hanging freely over the edge of the table. The distances AP and PB are $2h$ and h respectively, and B is at height h above a horizontal floor, as shown in the diagram.

The system is now released from rest and begins to move.

(a) Show that, until B hits the floor, the acceleration of A is $\frac{1}{4}(1 - 3\mu)g$.

(b) Find the speed of A immediately before B hits the floor.

After B hits the floor, A continues to move on the table and comes to rest before reaching P.

(c) Deduce the range of values of μ for which the motion described above is possible. [E]

43 A small sphere R, of mass 0.08 kg, moving with speed $1.5\,\text{m}\,\text{s}^{-1}$, collides directly with another small sphere S, of mass 0.12 kg, moving in the same direction with speed $1\,\text{m}\,\text{s}^{-1}$. Immediately after the collision R and S continue to move in the same direction with speeds $U\,\text{m}\,\text{s}^{-1}$ and $V\,\text{m}\,\text{s}^{-1}$ respectively.

Given that $U : V = 21:26$,

(a) show that $V = 1.3$,

(b) find the magnitude of the impulse, in N s, received by R as a result of the collision. [E]

44

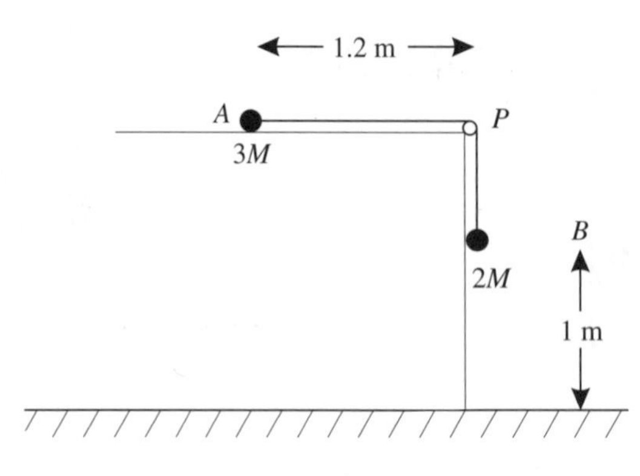

A particle A of mass $3M$ lies on a rough horizontal table. The particle is attached to a light inextensible string which passes over a small smooth pulley P fixed at the edge of the table. To the other end of the string is attached a particle B of mass $2M$, which hangs freely. AP is perpendicular to the edge of the table, and A, P and B are in the same vertical plane.
The system is released from rest with the string taut, when A is 1.2 m from the edge of the table and B is 1 m above the floor, as shown in the diagram. The particle B reaches the floor after 2 s, and does not rebound.

(a) Find the acceleration of A during the first two seconds.

(b) Find, to 2 decimal places, the coefficient of friction between A and the table.

(c) Determine, by calculation, whether A reaches P. [E]

45 A non-uniform thin straight rod AB has length $3d$ and mass $5m$. It is in equilibrium resting horizontally on supports at the points X and Y, where $AX = XY = YB = d$.
A particle of mass $2m$ is attached to the rod at B. Given that the rod is on the point of tilting about Y, find the distance of the centre of mass of the rod from A. [E]

Examination style paper

M1

Answer all questions **Time allowed 90 minutes**

Whenever a numerical value of g is required, take $g = 9.8\,\text{m s}^{-2}$.

1. Three forces $\mathbf{F}_1$, $\mathbf{F}_2$ and $\mathbf{F}_3$ act on a particle.

$$\mathbf{F}_1 = (-3\mathbf{i} + 2p\mathbf{j})\,\text{N},\ \mathbf{F}_2 = (p\mathbf{i} + 3q\mathbf{j})\,\text{N},\ \mathbf{F}_3 = (q\mathbf{i} - 7\mathbf{j})\,\text{N}$$

Given that the particle is in equilibrium, determine the value of p and the value of q.

(6 marks)

2. Two particles P and Q of masses kM and M are moving towards each other along the same horizontal straight line with speeds u and $2u$ respectively. After the impact Q is reduced to rest and the direction of motion of P is reversed.

(*a*) Find, in terms of k and u, the speed of P after the impact. **(3 marks)**

(*b*) State the condition k must satisfy. **(1 mark)**

(*c*) Calculate the magnitude of the impulse exerted by Q on P. **(2 marks)**

3. A car is moving along a straight road with uniform acceleration. The car passes point A with a speed of $15\,\text{m s}^{-1}$ and point B with a speed of $25\,\text{m s}^{-1}$. Given that the distance AB is 1000 m,

(*a*) find the time, in s, taken by the car to move from A to B. **(3 marks)**

M is the mid-point of AB.

(*b*) Find, in m s^{-1} to 1 decimal place, the speed with which the car passes M. **(4 marks)**

4. In this question $\mathbf{i}$ and $\mathbf{j}$ are unit vectors due east and due north respectively.

At 2 p.m. the position vector relative to a lighthouse of ship A is $(3\mathbf{i} + \mathbf{j})$ km and its velocity is $(-\mathbf{i} + 5\mathbf{j})\,\text{km h}^{-1}$. At the same time ship B has position vector $(-\mathbf{i} + 4\mathbf{j})$ km relative to the lighthouse and velocity $(2\mathbf{i} + 3\mathbf{j})\,\text{km h}^{-1}$. Find, after t h,

(*a*) the position vector of A relative to the lighthouse, **(2 marks)**

(*b*) the position vector of B relative to the lighthouse, **(2 marks)**

(*c*) the position vector of A relative to B. **(2 marks)**
(*d*) Find the time when A is due north of B. **(3 marks)**

5. Two plasterers Bill and Dave of mass 45 kg and 65 kg respectively stand on a uniform plank $ABCDEF$ of length 4 m and mass 40 kg. The plank is supported at C and E where C is 1 m from A and E is 1 m from F. Plasterer Bill stands at B which is 0.5 m from A and plasterer Dave stands at D which is 1.5 m from F.
(*a*) Draw a diagram showing the above information. **(2 marks)**
(*b*) Find the reaction at each support, giving your answers in N to 1 decimal place.

(7 marks)

6.

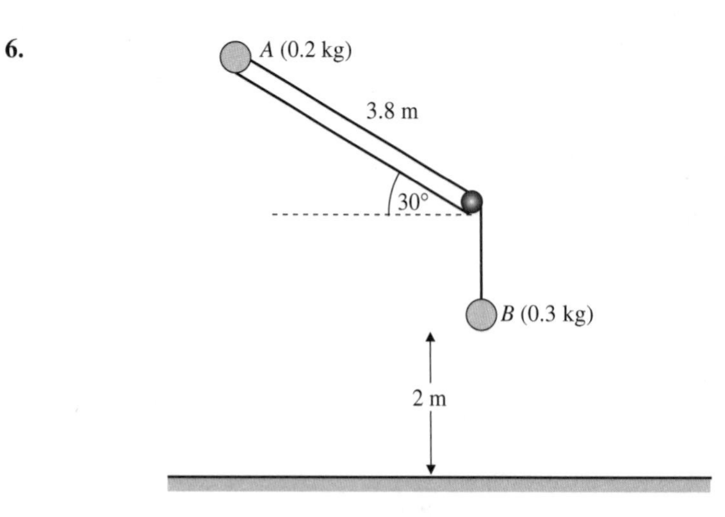

Figure 1

Figure 1 shows a particle A of mass 0.2 kg held at rest on a smooth roof inclined at 30° to the horizontal. Particle A is attached to one end of a light inextensible string which passes over a smooth pulley at the edge of the roof. A particle B of mass 0.3 kg is attached to the other end of the string and hangs freely at rest. Particle A is 3.8 m from the pulley and particle B is 2 m above the horizontal ground.
The system is released from rest.
(*a*) Find the acceleration of each particle while the string remains taut.

(6 marks)

(*b*) Find the speed with which B strikes the floor.

(2 marks)

Assuming that B is brought to rest when it strikes the floor,
(*c*) find the speed with which A reaches the pulley.

(4 marks)

In this problem several mathematical models have been used. Identify three of these and briefly describe the assumptions which have been made in using these models.

(2 marks)

7.

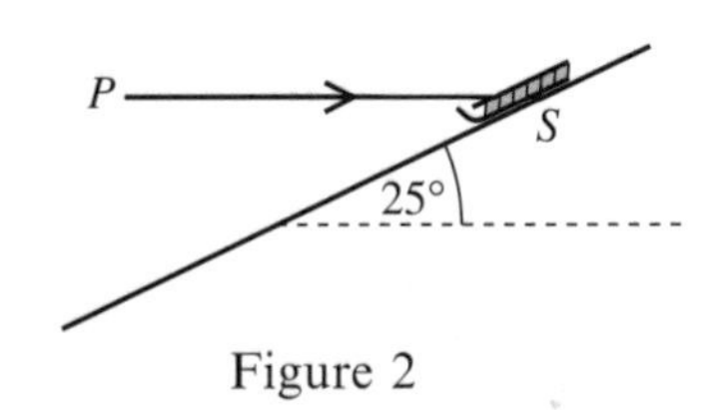

Figure 2

Figure 2 shows a sledge S of mass 20 kg on an icy straight track inclined at 25° to the horizontal. The coefficient of friction μ between S and the track is 0.2. The position of S is maintained by a horizontal force of P newtons which acts in the vertical plane containing the track. The sledge S is in limiting equilibrium and on the point of moving down the track. Modelling the sledge as a particle,

(*a*) obtain, to 3 significant figures, the value of P. **(7 marks)**

The force P is removed and S moves down the track from rest. Given that μ is still 0.2,

(*b*) find the initial acceleration of S to 2 significant figures.

(4 marks)

8. At a point P a car is moving with speed $20\,\text{m}\,\text{s}^{-1}$. It then accelerates with a uniform acceleration of $3\,\text{m}\,\text{s}^{-2}$ until it reaches a maximum speed of $50\,\text{m}\,\text{s}^{-1}$. It then decelerates at $4\,\text{m}\,\text{s}^{-2}$ until it comes to rest at Q.

(*a*) Sketch a speed–time graph for the motion of the car from P to Q.

(3 marks)

(*b*) Find the total time taken to travel from P to Q. **(4 marks)**

(*c*) Find the distance PQ. **(5 marks)**

Answers

Note on accuracy: Unless otherwise stated all numerical answers are given to 3 significant figures. When calculating answers the maximum number of figures possible should be used before the final answer is rounded to 3 significant figures. Rounding intermediate answers can produce an inaccurate final answer.

Edexcel accepts no responsibility whatsoever for the accuracy or method of working in the answers given for examination questions.

Exercise 1A

1 *Car:* cost of petrol and of parking.
Assumptions: include mileage, price of petrol, miles per gallon, availability of convenient car park.
Coach: fares, cost to and from coach station.
Assumption: coach times convenient.
Train: fares, cost to and from station, *Assume:* convenient train available.
Possible combinations include car to bus or train station, car to free car park on outskirts plus underground train to city centre.
Refinements: could include consideration of time factor and its implications for number of meals required. Also consideration of time of day and possible delays.

2 *Relevant costs:* gas, electricity, telephone, water, Council Tax, food, transport, mortgage or rent.
Assumption: costs will be $\frac{1}{4}$ of last year's bills, and there are no unforeseen costs.
Refinements: could include more accurate estimate taking into account a particular 3 month period, and some allowance for emergencies based on past experience.

3 *Relevant facts:* length of three routes, numbers of: traffic lights, roundabouts, right turns, railway crossings.
Assumption: he drives at a constant speed, with a fixed delay for each of traffic lights, roundabouts etc.
Refinements: could include trying to obtain more accurate estimates for the delay based on trying each route for a week.

4 *Relevant facts:* distance covered in a year, cost of petrol per litre, fuel efficiency of car (km per litre), depreciation of car, servicing cost of car, M.O.T. test cost, insurance cost, motoring organisation (AA or similar).
Assumptions: distance covered same as last year, price of petrol is average from last year, car less efficient than last year due to age, depreciation taken as average for particular model, insurance less due to increase in no claims bonus, servicing costs as last year.
Refinements: include some allowance for unforeseen factors based on past experience. Carry out a trial to obtain better estimate of fuel efficiency in km per litre. Obtain better estimate of depreciation.

5 *Relevant factors:* costs of: hiring rooms, equipment, providing transport from nearest station, publicity, telephone, postage, printing and secretarial help.

Assumptions: at least 100 people will attend; 50% will need transport from station; delegates will divide themselves equally between various groups.
Refinements: could obtain indication of number who will attend by using a circular. This could also give a better estimate of number requiring transport.

6 *Relevant factors:* cost of a typical shopping list, cost of transport to new supermarket.
Assumptions: prices at shop and supermarket remain basically the same; cost of transport does not vary depending on time spent on trip.
Refinements: take into account savings due to bulk purchases and own brands.

7 (a) *Relevant factors:* costs of: flying, hiring car, transport to and from airport.
(b) *Relevant factors:* costs of boat, of taking car, or cost of hiring and getting to boat.
(c) *Relevant factors:* cost of catamaran, of taking car, or cost of hiring and getting to port.
Assumptions: these are the only costs involved, and are independent of time and season.
Refinements: the travelling times for (a), (b) and (c) are different so the cost of meals needs to be taken into account. The costs of (a), (b) and (c) will vary according to time of year and time of day.

Exercise 1B

1 (a) A particle of mass 1 kg attached to one end of a light inextensible string the other end of which is attached to a fixed point. The particle moves in a vertical circle under constant gravity only.
(b) A particle of mass 0.5 kg is released from rest on a smooth plane inclined at 20° to the horizontal.
(c) A particle moves on a horizontal plane and collides with a similar particle at rest on the plane.
(d) A uniform rod of length 4 m can rotate in a vertical plane about a horizontal axis through its centre. A particle of mass 25 kg is attached at one end. A second particle of mass 20 kg is attached to the rod at the point P so that the rod rests in a horizontal position.
(e) A uniform rod rests on a rough horizontal plane with its upper end against a rough vertical wall. The rod makes an angle α with the horizontal. A particle is attached at the upper end. Find the greatest value of α for which the rod does not slip.

2 (a) This is a good model as a cricket ball is small and the air resistance is likely to be small. The variation of gravity is not great for the distances involved.
Refinements: The model may be refined by taking into account air resistance and the dependence of gravity on height.
(b) This is a good model as the puck is small and ice offers little frictional resistance to motion.
Refinements: The model may be refined by taking into account friction and also air resistance.
(c) This is a reasonable model as the size of the spheres is small compared with the other lengths involved and the mass of the rod is small compared with the other masses involved.
Refinements: The model can be refined by taking into account the shape of the spheres, the mass of the rod, the variation of gravity, and air resistance.
(d) This is a reasonable model as the size of the bucket is likely to be small compared with the length of the rope. The assumption that the string is light is valid as the weight of the bucket is likely to be much greater than that of the rope. The extension of the rope is also likely to be small so that the string may be taken as inextensible. The assumption that the pulley is smooth (its bearings are completely without friction) is likely to be reasonable if the equipment is well maintained.
Refinements: Many refinements are possible including replacement of the particle by a mass of finite size, replacement of the inextensible string by one which does extend. In addition the

friction due to the bearings of the pulley and the mass of the rope could be taken into account at a later stage.

(e) This is not a very good model as the assumption that the pole may be replaced by a rigid rod neglects a fundamental property of the pole – its flexibility.

Exercise 2A

1 (a) 17 km, 13 km (b) 17 km, 13 km (c) 17 km, 12.2 km (d) 17 km, 16.0 km (e) 37 km, 28.8 km

2 (a) 7 km, 6.57 km (b) 7 km, 6.48 km (c) 7 km, 6.57 km (d) 7 km, 6.74 km (e) 9 km, 7.33 km

3 (a) 9.43 km, 032° (b) 9.43 km, 058° (c) 18.9 km, 032°

4 (a) 9.22 km, 091° (b) 9.22 km, 099° (c) 18.4 km, 099°

5 (a) 8.4 km, 155° (b) 7.7 km, 150° (c) 10.7 km, 167°

6 (a) $1.12\,\text{m s}^{-1}$, 26.6° (b) $1.58\,\text{m s}^{-1}$, 18.4° (c) $2.24\,\text{m s}^{-1}$, 63.4°

Exercise 2B

1 (a) $\frac{1}{2}\mathbf{a}$ (b) $-\frac{1}{2}\mathbf{b}$ (c) $\mathbf{b}-\mathbf{a}$ (d) $\mathbf{b}$ (e) $-\mathbf{b}$ (f) $-\frac{1}{2}\mathbf{b}$ (g) $\frac{1}{2}(\mathbf{b}-\mathbf{a})$ (h) $\mathbf{a}+\mathbf{b}$ (i) $-(\mathbf{a}+\mathbf{b})$ (j) $\frac{1}{2}\mathbf{a}+\mathbf{b}$.

2 (a) $\mathbf{b}$ (b) $-\mathbf{a}$ (c) $\mathbf{a}+\frac{1}{2}\mathbf{b}$ (d) $-\frac{1}{2}\mathbf{a}+\mathbf{b}$ (e) $\frac{1}{2}\mathbf{a}-\mathbf{b}$ (f) $-\frac{1}{2}\mathbf{a}$ (g) $\frac{1}{2}(\mathbf{a}-\mathbf{b})$ (h) $\frac{1}{2}(\mathbf{b}-\mathbf{a})$ (i) $\mathbf{b}-\mathbf{a}$ (j) $\frac{1}{2}\mathbf{b}-\mathbf{a}$

3 (a) $\mathbf{a}+\mathbf{b}$ (b) $-\mathbf{a}+\mathbf{b}$ (c) $-\frac{1}{2}\mathbf{a}+\mathbf{b}$ (d) $-\frac{1}{2}\mathbf{a}-\mathbf{b}$ (e) $\frac{1}{2}\mathbf{a}+\mathbf{b}$ (f) $\frac{1}{2}(\mathbf{a}+\mathbf{b})$ (g) $-\mathbf{a}-\frac{1}{2}\mathbf{b}$ (h) $-\frac{1}{2}(\mathbf{a}+\mathbf{b})$ (i) $-\frac{1}{2}\mathbf{b}$ (j) $-\mathbf{b}$

4 (a) $\overrightarrow{AO}$ (b) $\overrightarrow{OE}$ or $\overrightarrow{BO}$ (c) $\overrightarrow{OC}$ (d) $\overrightarrow{OD}$ (e) $\overrightarrow{DO}$.

5 (a) $-\mathbf{a}$ (b) $\mathbf{b}$ (c) $\mathbf{c}$ (d) $\mathbf{a}+\mathbf{b}$ (e) $\mathbf{a}+\mathbf{b}+\mathbf{c}$ (f) $\mathbf{b}+\mathbf{c}$ (g) $-\mathbf{a}+\mathbf{c}$ (h) $\mathbf{a}-\mathbf{c}$ (i) $-\mathbf{a}-\mathbf{b}$ (j) $-\mathbf{b}-\mathbf{c}$

Exercise 2C

1 (a) $3\mathbf{i}+\mathbf{j}$ (b) $5\mathbf{i}$ (c) $4\mathbf{i}+3\mathbf{j}$ (d) $-3\mathbf{i}+4\mathbf{j}$ (e) $5\mathbf{j}$

2 (a) $-\frac{1}{3}$ (b) $-\frac{1}{2}$

3 (a) $\frac{1}{2}$ (b) 1

4 (a) (i) 13 (ii) 8.60 (iii) 8.60 (iv) 9.43 (v) 12.2 (b) (i) 67.4° (ii) 54.5° (iii) −54.5° (iv) −122° (v) −55°.

5 (a) 3.61, 56.3° (b) 3.61, −123.7° (c) 7.21, 56.3° (d) 1.80, 56.3° (e) 18.0, −123.7.

6 (a) 3, 0° (b) 2.24, 117° (c) 4.12, 14° (d) 5.10, 101° (e) 6.40, −38.7°

7 (a) $\frac{1}{\sqrt{2}}(\mathbf{i}+\mathbf{j})$ (b) $\frac{1}{\sqrt{2}}(\mathbf{i}-\mathbf{j})$ (c) $\frac{1}{5}(3\mathbf{i}-4\mathbf{j})$ (d) $\frac{1}{5}(-3\mathbf{i}+4\mathbf{j})$ (e) $\frac{1}{5}(3\mathbf{i}-4\mathbf{j})$

8 (a) 1.73, 1 (b) 2.60, 1.5 (c) 5.20, 3 (d) 8.66, 5

9 (i) 5, 8.66 (ii) −5, 8.66 (iii) −8.66, 5 (iv) −5, −8.66 (v) 5, −8.66

10 (i) (a) 3,0; 0,4 (b) 3, 4 (c) 5
(ii) (a) 3,0; 2, 3.46 (b) 5, 3.46 (c) 6.08
(iii) (a) 2.60, 1.5; 2, 3.46 (b) 4.60, 4.96 (c) 6.76
(iv) (a) 4.33, 2.5; 4, 6.93 (b) 8.33, 9.43 (c) 12.6
(v) (a) 4.33, 2.5; −6.93, 4 (b) −2.60, 6.5 (c) 7

Exercise 2D

1 (a) $2\mathbf{i}+3\mathbf{j}$ (b) $-4\mathbf{j}$ (c) $2\mathbf{i}+7\mathbf{j}$ (d) $3\mathbf{i}+3\mathbf{j}$ (e) $-\mathbf{i}+3\mathbf{j}$

2 (a)

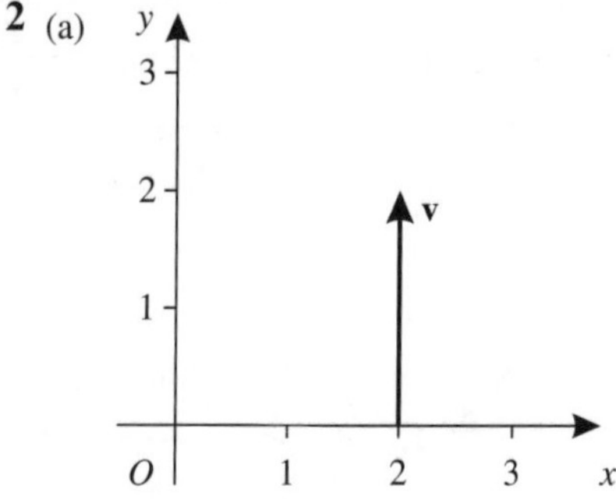

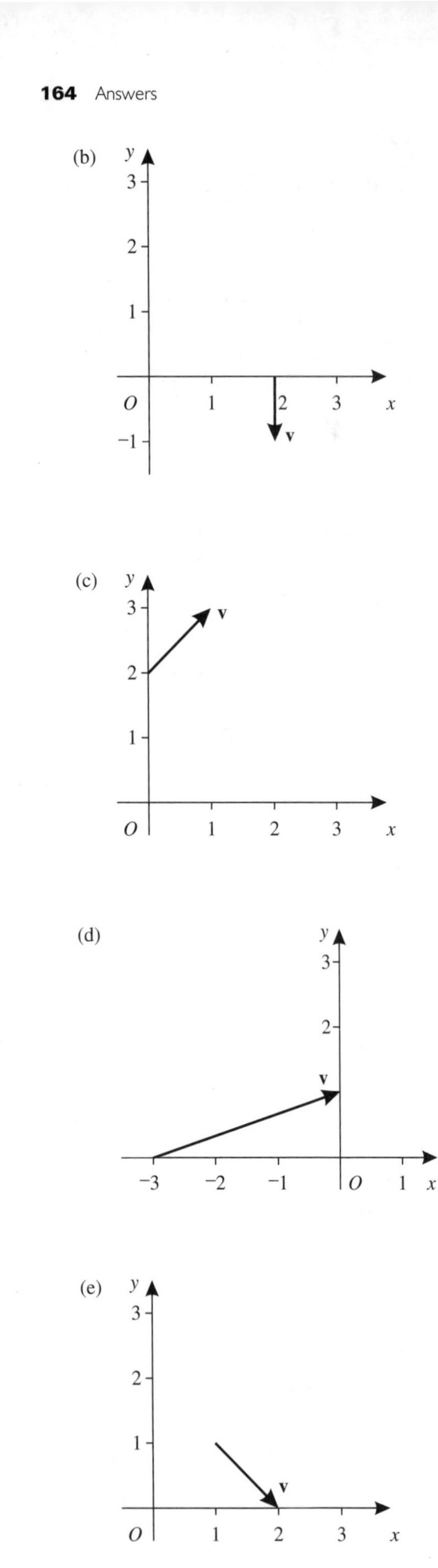

3 (a) $2\mathbf{i} + 2\mathbf{j}$ (b) $2\mathbf{i} - \mathbf{j}$ (c) $\mathbf{i} + 3\mathbf{j}$ (d) $\mathbf{j}$ (e) $2\mathbf{i}$

4 (a) $(\mathbf{i} + 2\mathbf{j})$ (b) $(-\mathbf{i} + 2\mathbf{j})$ (c) $(\mathbf{i} + 2\mathbf{j})$ (d) $(\mathbf{i} - \mathbf{j})$ (e) $(\mathbf{i} + 2\mathbf{j})$

5 (a) $3\mathbf{i}$; 3 (b) $3\mathbf{i} + 4\mathbf{j}$; 5 (c) $-3\mathbf{i} + 4\mathbf{j}$; 5 (d) $3\mathbf{i} + 4\mathbf{j}$; 5 (e) $4\mathbf{i} + 3\mathbf{j}$; $5\,\text{m s}^{-1}$

6 (a) $5\mathbf{i}$; $5\,\text{m s}^{-1}$ (b) $(3\mathbf{i} + 4\mathbf{j})$; $5\,\text{m s}^{-1}$
(c) $(5\mathbf{i} + 12\mathbf{j})$; $13\,\text{m s}^{-1}$
(d) $(5\mathbf{i} - 12\mathbf{j})$; $13\,\text{m s}^{-1}$
(e) $(8\mathbf{i} + 15\mathbf{j})$; $17\,\text{m s}^{-1}$

7 (a) $\frac{1}{2}\mathbf{i}$ (b) $(-\mathbf{i} + 2\mathbf{j})$ (c) $(-\mathbf{i} - \mathbf{j})$ (d) $(-4\mathbf{i} + \mathbf{j})$ (e) $(3\mathbf{i} - 5\mathbf{j})$

8 $\frac{1}{5}(3\mathbf{i} - 4\mathbf{j})$; 1

9 $20\mathbf{i} - 23\mathbf{j}$; 30 m

10 $11\mathbf{i} + 12\mathbf{j}$; $\sqrt{265}\,\text{m s}^{-1}$

11 (a) $\mathbf{r}_S = 600\mathbf{j} + (7\mathbf{i} + 8\mathbf{j})t$
$\mathbf{r}_B = 120\mathbf{j} + (7\mathbf{i} + 24\mathbf{j})t$
(b) 30 s after noon, $(210\mathbf{i} + 840\mathbf{j})$ m
(c) 330 m

12 (c) $\sqrt{17}$ km

Exercise 3A

1 $15\,\text{m s}^{-1}$ **2** $2.25\,\text{m s}^{-2}$
3 $17\,\text{m s}^{-1}$ **4** 4.4 s
5 18 m **6** $3.25\,\text{m s}^{-1}$
7 1 s **8** 2 s, 28 m
9 45 m **10** $2.5\,\text{m s}^{-2}$
11 150 m, $1.4\,\text{m s}^{-2}$
12 $10\,\text{m s}^{-1}$ **13** 227 m, $7.14\,\text{m s}^{-1}$
14 $2\,\text{m s}^{-2}$, 21 m **15** $3\frac{1}{3}\,\text{m s}^{-2}$, $13\frac{1}{3}\,\text{m s}^{-1}$

Exercise 3B

1 $4.43\,\text{m s}^{-1}$ **2** 2.02 s **3** 11.5 m
4 (a) 5.10 m (b) 2.04 s
5 (a) $25.8\,\text{m s}^{-1}$ (b) 5.18 s
6 $10.8\,\text{m s}^{-1}$ **7** 2.4 m **8** $15.6\,\text{m s}^{-1}$, 3.12 m
9 2.56 s **10** 23.0 m
11 2.47 s. Air resistance can be neglected; acceleration due to gravity is constant. Answer is slightly too small.

Exercise 3C

1 (a) $2.5\,\text{m}\,\text{s}^{-2}$ (b) $1.25\,\text{m}\,\text{s}^{-2}$ (c) 45 m

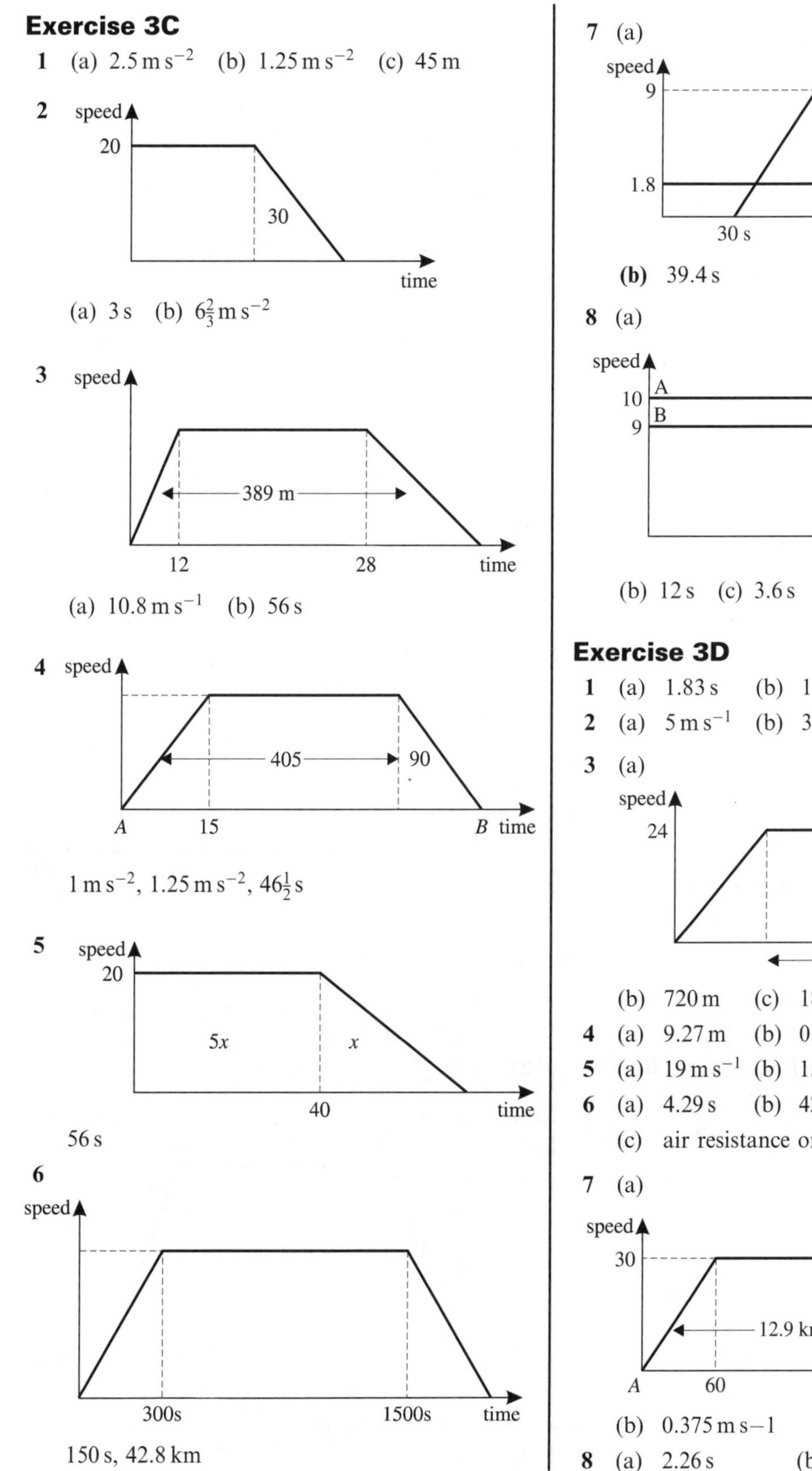

2 (a) 3 s (b) $6\frac{2}{3}\,\text{m}\,\text{s}^{-2}$

3 (a) $10.8\,\text{m}\,\text{s}^{-1}$ (b) 56 s

4 $1\,\text{m}\,\text{s}^{-2}$, $1.25\,\text{m}\,\text{s}^{-2}$, $46\frac{1}{2}$ s

5 56 s

6 150 s, 42.8 km

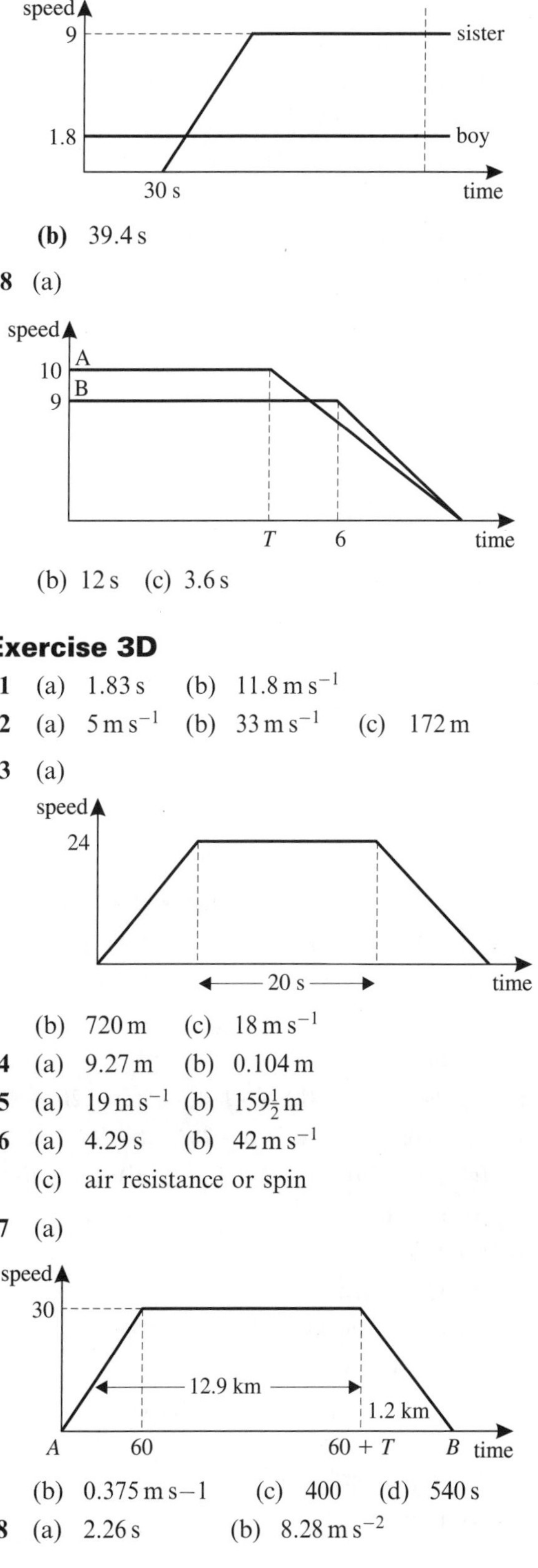

7 (a) (b) 39.4 s

8 (a) (b) 12 s (c) 3.6 s

Exercise 3D

1 (a) 1.83 s (b) $11.8\,\text{m}\,\text{s}^{-1}$

2 (a) $5\,\text{m}\,\text{s}^{-1}$ (b) $33\,\text{m}\,\text{s}^{-1}$ (c) 172 m

3 (a) (b) 720 m (c) $18\,\text{m}\,\text{s}^{-1}$

4 (a) 9.27 m (b) 0.104 m

5 (a) $19\,\text{m}\,\text{s}^{-1}$ (b) $159\frac{1}{2}$ m

6 (a) 4.29 s (b) $42\,\text{m}\,\text{s}^{-1}$
(c) air resistance or spin

7 (a) (b) 0.375 m s–1 (c) 400 (d) 540 s

8 (a) 2.26 s (b) $8.28\,\text{m}\,\text{s}^{-2}$

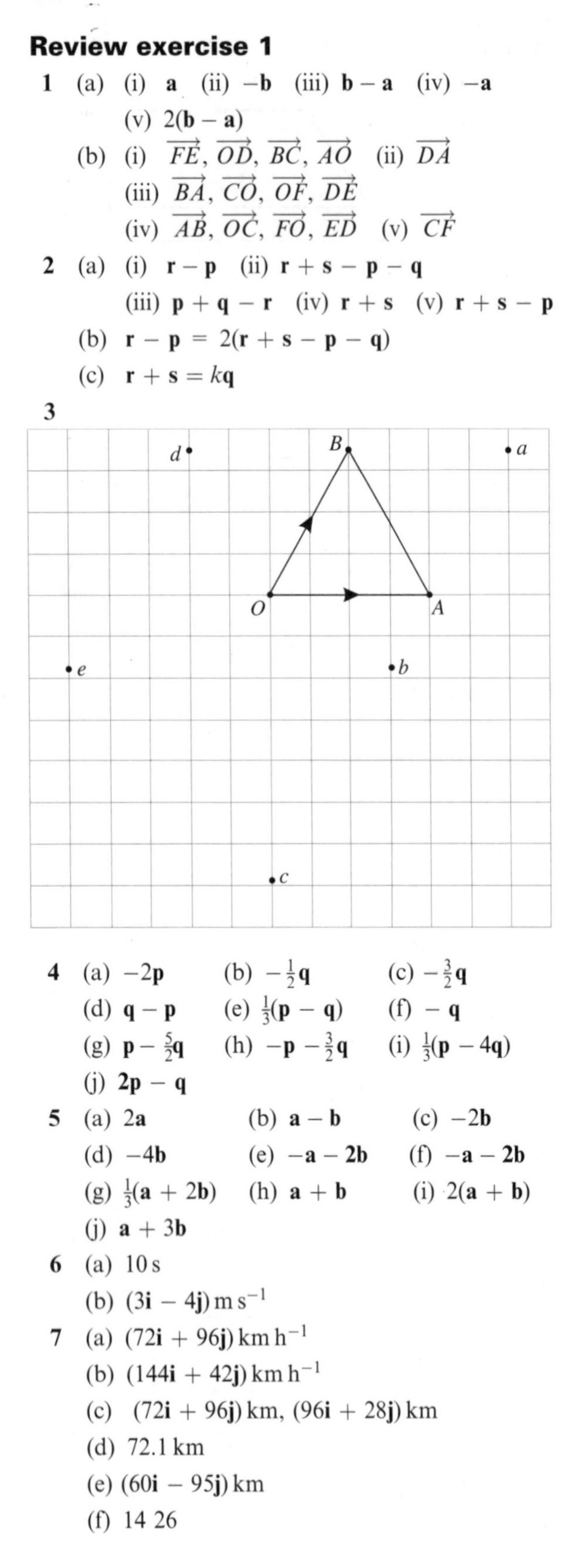

Review exercise 1

1 (a) (i) $\mathbf{a}$ (ii) $-\mathbf{b}$ (iii) $\mathbf{b}-\mathbf{a}$ (iv) $-\mathbf{a}$
(v) $2(\mathbf{b}-\mathbf{a})$
(b) (i) $\overrightarrow{FE}$, $\overrightarrow{OD}$, $\overrightarrow{BC}$, $\overrightarrow{AO}$ (ii) $\overrightarrow{DA}$
(iii) $\overrightarrow{BA}$, $\overrightarrow{CO}$, $\overrightarrow{OF}$, $\overrightarrow{DE}$
(iv) $\overrightarrow{AB}$, $\overrightarrow{OC}$, $\overrightarrow{FO}$, $\overrightarrow{ED}$ (v) $\overrightarrow{CF}$

2 (a) (i) $\mathbf{r}-\mathbf{p}$ (ii) $\mathbf{r}+\mathbf{s}-\mathbf{p}-\mathbf{q}$
(iii) $\mathbf{p}+\mathbf{q}-\mathbf{r}$ (iv) $\mathbf{r}+\mathbf{s}$ (v) $\mathbf{r}+\mathbf{s}-\mathbf{p}$
(b) $\mathbf{r}-\mathbf{p} = 2(\mathbf{r}+\mathbf{s}-\mathbf{p}-\mathbf{q})$
(c) $\mathbf{r}+\mathbf{s} = k\mathbf{q}$

3

4 (a) $-2\mathbf{p}$ (b) $-\frac{1}{2}\mathbf{q}$ (c) $-\frac{3}{2}\mathbf{q}$
(d) $\mathbf{q}-\mathbf{p}$ (e) $\frac{1}{3}(\mathbf{p}-\mathbf{q})$ (f) $-\mathbf{q}$
(g) $\mathbf{p}-\frac{5}{2}\mathbf{q}$ (h) $-\mathbf{p}-\frac{3}{2}\mathbf{q}$ (i) $\frac{1}{3}(\mathbf{p}-4\mathbf{q})$
(j) $2\mathbf{p}-\mathbf{q}$

5 (a) $2\mathbf{a}$ (b) $\mathbf{a}-\mathbf{b}$ (c) $-2\mathbf{b}$
(d) $-4\mathbf{b}$ (e) $-\mathbf{a}-2\mathbf{b}$ (f) $-\mathbf{a}-2\mathbf{b}$
(g) $\frac{1}{3}(\mathbf{a}+2\mathbf{b})$ (h) $\mathbf{a}+\mathbf{b}$ (i) $2(\mathbf{a}+\mathbf{b})$
(j) $\mathbf{a}+3\mathbf{b}$

6 (a) 10 s
(b) $(3\mathbf{i}-4\mathbf{j})\,\text{m s}^{-1}$

7 (a) $(72\mathbf{i}+96\mathbf{j})\,\text{km h}^{-1}$
(b) $(144\mathbf{i}+42\mathbf{j})\,\text{km h}^{-1}$
(c) $(72\mathbf{i}+96\mathbf{j})$ km, $(96\mathbf{i}+28\mathbf{j})$ km
(d) 72.1 km
(e) $(60\mathbf{i}-95\mathbf{j})$ km
(f) 14 26

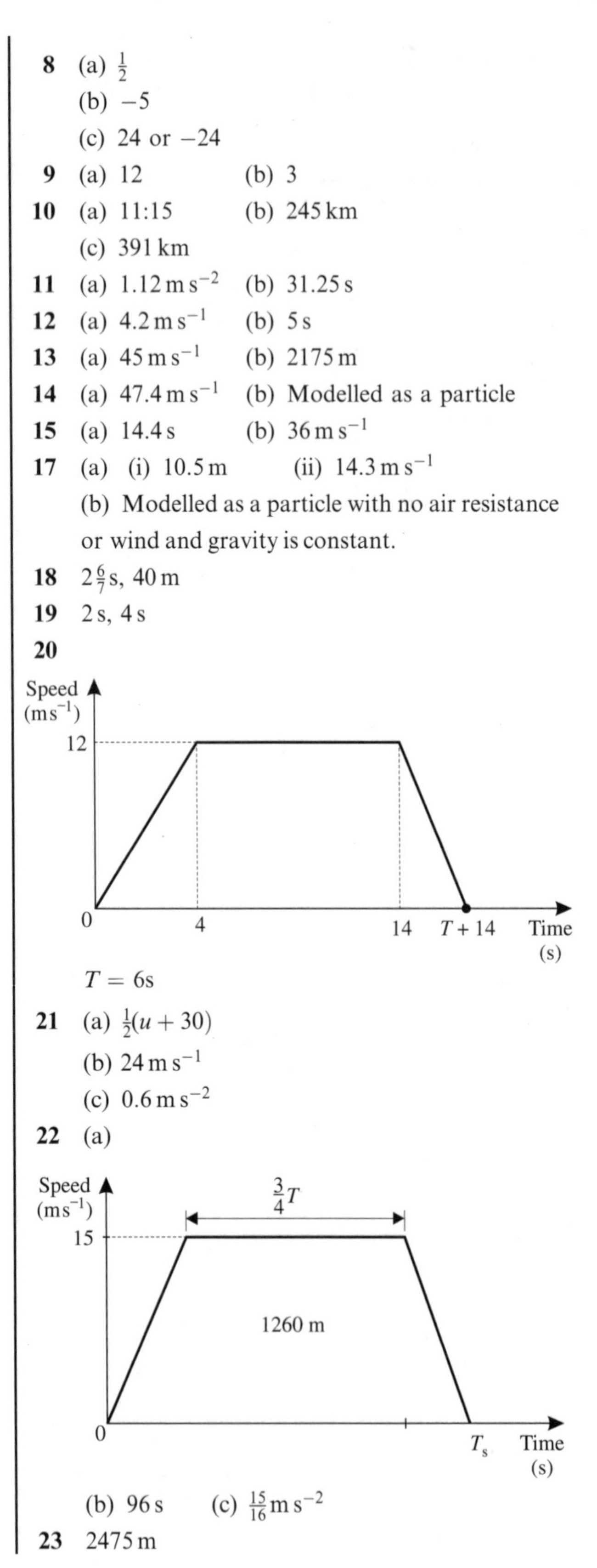

8 (a) $\frac{1}{2}$
(b) -5
(c) 24 or -24

9 (a) 12 (b) 3

10 (a) 11:15 (b) 245 km
(c) 391 km

11 (a) $1.12\,\text{m s}^{-2}$ (b) 31.25 s

12 (a) $4.2\,\text{m s}^{-1}$ (b) 5 s

13 (a) $45\,\text{m s}^{-1}$ (b) 2175 m

14 (a) $47.4\,\text{m s}^{-1}$ (b) Modelled as a particle

15 (a) 14.4 s (b) $36\,\text{m s}^{-1}$

17 (a) (i) 10.5 m (ii) $14.3\,\text{m s}^{-1}$
(b) Modelled as a particle with no air resistance or wind and gravity is constant.

18 $2\frac{6}{7}$ s, 40 m

19 2 s, 4 s

20

$T = 6$s

21 (a) $\frac{1}{2}(u+30)$
(b) $24\,\text{m s}^{-1}$
(c) $0.6\,\text{m s}^{-2}$

22 (a)

(b) 96 s (c) $\frac{15}{16}\,\text{m s}^{-2}$

23 2475 m

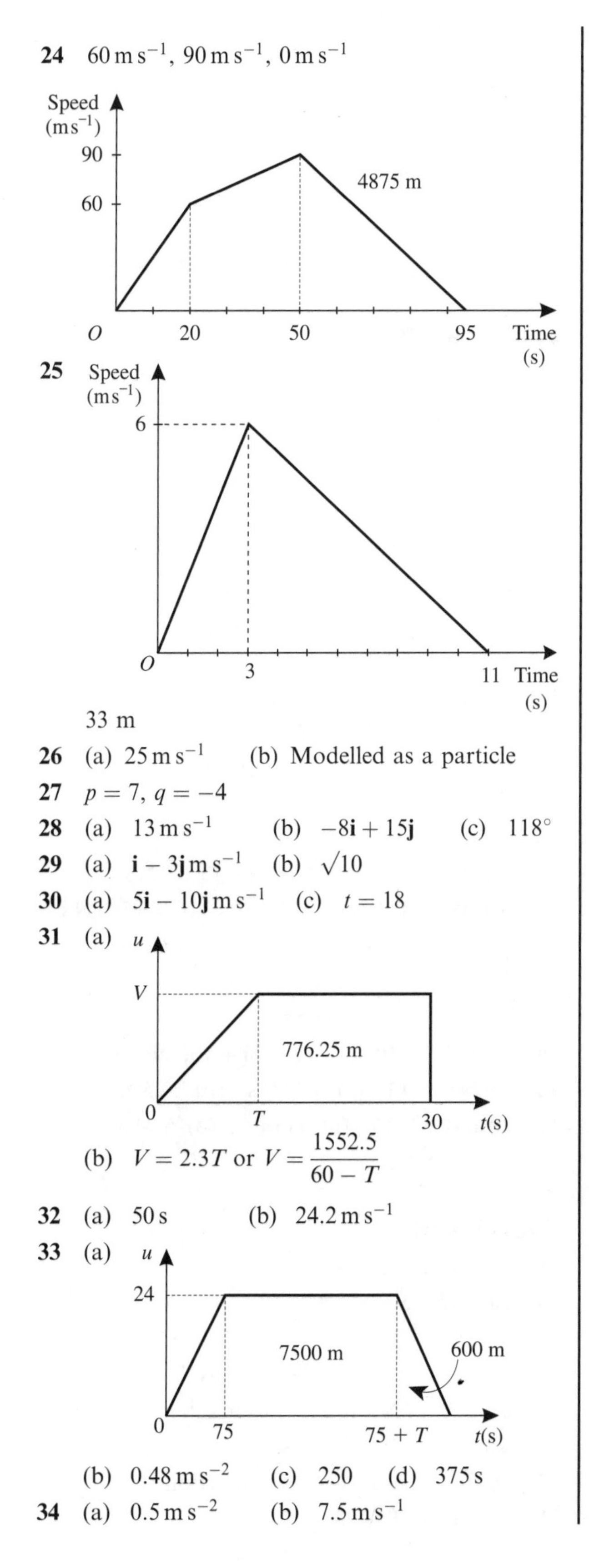

24 $60\,\text{m s}^{-1}$, $90\,\text{m s}^{-1}$, $0\,\text{m s}^{-1}$

25

33 m

26 (a) $25\,\text{m s}^{-1}$ (b) Modelled as a particle

27 $p = 7$, $q = -4$

28 (a) $13\,\text{m s}^{-1}$ (b) $-8\mathbf{i} + 15\mathbf{j}$ (c) $118°$

29 (a) $\mathbf{i} - 3\mathbf{j}\,\text{m s}^{-1}$ (b) $\sqrt{10}$

30 (a) $5\mathbf{i} - 10\mathbf{j}\,\text{m s}^{-1}$ (c) $t = 18$

31 (a)

(b) $V = 2.3T$ or $V = \dfrac{1552.5}{60 - T}$

32 (a) $50\,\text{s}$ (b) $24.2\,\text{m s}^{-1}$

33 (a)

(b) $0.48\,\text{m s}^{-2}$ (c) 250 (d) 375 s

34 (a) $0.5\,\text{m s}^{-2}$ (b) $7.5\,\text{m s}^{-1}$

Exercise 4A

1 (a) 5 N, 36.9° (b) 9.43 N, 58.0°
(c) 9.43 N, 32.0° (d) 28.3 N, 32.0°
(e) 41 N, 77.3°

2 (a) 6.08 N, 25.3° (b) 6.08 N, 34.7°
(c) 8.66 N, 30° (d) 11.1 N, 68.9°
(e) 7 N, 38.2°

3 (a) 9.98 N, 56.6° (b) 9.80 N, 48.4°
(c) 14.8 N, 47.7° (d) 14.4 N, 51.2°
(e) 14.5 N, 45°

4 (a) $(10\mathbf{i} + 11\mathbf{j})$ N (b) $(5\mathbf{i} + 10\mathbf{j})$ N
(c) $7\mathbf{i}$ N (d) $-6\mathbf{j}$ N (e) 0

5 (a) 14.9 N, 47.7° (b) 11.2 N, 63.4°
(c) 7 N, 0° (d) 6 N, 90° (e) 0

6 (a) 8.4 N, 95° (b) 3.1 N, 124°
(c) 3.5 N, 26°

Exercise 4B

1 (a) 3.46 N, 2 N (b) 3 N, 5.20 N
(c) −3 N, 5.20 N (d) −6.93 N, 4 N
(e) −5 N, −8.66 N

2 (a) 1.88 N, 0.684 N (b) 8.66 N, 5 N
(c) 12.9 N, 15.3 N

3 (a) 1 N, 1.73 N (b) 3.06 N, 2.57 N
(c) 9.66 N, 2.59 N

4 (a) 2.24 N, −49.4° (b) 2.60 N, −8.33°
(c) 3.79 N, −75.3°

5 (a) 3.14 N, 42.5° (b) 4.40 N, 37.3°
(c) 27.5 N, 65.6°

Exercise 4C

1 (i) 5 N, 53.1° (ii) 9.43 N, 58.0°
(iii) 13 N, 67.4°

2 (i) 0.266 N, 6.36 N
(ii) 0.732 N, 10.4 N
(iii) 4.93 N, 18.9 N

3 (a) (i) $x = -5$, $y = -8$ (b) $x = 1$, $y = 7$
(c) $x = -5$, $y = 6$ (d) $x = 5$, $y = 2$
(e) $x = -1$, $y = -1$

4 $p = 2$, $q = -6$

Exercise 4D

1 (a) 1.96 N (b) 0.5 kg (c) 9.8 N (d) 5 N

2 (a) 8 N (b) 21.6 N (c) 6 N (d) 1.02 kg

3 (a) 24.5 N (b) 63 N (c) 2.74 N

4 (a) 5.1 N down the line of greatest slope
(b) 0.2 N down the line of greatest slope
(c) 4.7 N up the line of greatest slope
(d) 4.8 N up the line of greatest slope
(e) 3 N down the line of greatest slope

5 (a) 17.3 N, 39 N (b) 8.66 N, 4.8 N
(c) 17.3 N, 2 N

6 (a) 0.406 N (b) 0.498 N

7 (a) 14.1 N, 14.1 N (b) 12.8 N, 12.9 N
(c) 8.82 N, 8.84 N
If α and β are not nearly equal, the peg-bag would slip on the washing line.

8 (a) 2.83 N, 2.83 N (b) 2.53 N, 2.99 N

Exercise 4E

1 (a) 2 N (b) 0.5 (c) 20 N

2 (a) 21.7 N (b) 28.9 N

3 (a) 36.5 N, 0.593 (b) 34.6 N, 0.723

4 (a) 6.70 N up the plane (b) 0.36

5 (a) 11.9 N (b) 3.00 N

6 (a) 43 N (b) 3.67N

Exercise 5A

1 $6\,\text{m}\,\text{s}^{-2}$ **2** 30 N **3** 20 N **4** $5\,\text{m}\,\text{s}^{-2}$

5 10 N **6** 4 kg **7** $(\frac{4}{5}\mathbf{i} + \frac{6}{5}\mathbf{j})\,\text{m}\,\text{s}^{-2}$

8 $(8\mathbf{i} - 4\mathbf{j})$ N **9** $(4\mathbf{i} + \mathbf{j})\,\text{m}\,\text{s}^{-2}$ **10** 26 N

11 $3.75\,\text{m}\,\text{s}^{-2}$, 2100 N **12** $1.5\,\text{m}\,\text{s}^{-2}$, 3 m

13 390 N **14** 4, −2 **15** $0.64\,\text{m}\,\text{s}^{-2}$

16 2000 N **17** 1500 N **18** 1470 N

19 10.8 N, ($\sqrt{117}$ N) **20** 0.0638

Exercise 5B

1 2.67 N **2** 0.372 N **3** 510 N, 510 N

4 1330 kg **5** $3.35\,\text{m}\,\text{s}^{-2}$ **6** 62.1 kg

7 30° **8** $6.93\,\text{m}\,\text{s}^{-2}$ **9** $2.49\,\text{m}\,\text{s}^{-2}$

10 $0.776\,\text{m}\,\text{s}^{-2}$ **11** $7.67\,\text{m}\,\text{s}^{-1}$ **12** 0.201

13 0.153 **14** 0.241 **15** 0.0827, $1.97\,\text{m}\,\text{s}^{-2}$

16 $1.97\,\text{m}\,\text{s}^{-2}$ **17** $2.35\,\text{m}\,\text{s}^{-2}$, 1.45 m

18 $5.20\,\text{m}\,\text{s}^{-2}$, $10.4\,\text{m}\,\text{s}^{-1}$

Exercise 5C

1 150 N, $0.25\,\text{m}\,\text{s}^{-2}$ **2** 786 N

3 4.44 N, $1.31\,\text{m}\,\text{s}^{-2}$ **4** 1820 N, 650 N

5 $966\frac{2}{3}$ N, $483\frac{1}{3}$ N, $733\frac{1}{3}$ N **6** 2500 N, 1500 N

Exercise 5D

1 $3.27\,\text{m}\,\text{s}^{-2}$, 26.1 N **2** $2.67\,\text{m}\,\text{s}^{-2}$, 99.8 N

3 $\frac{g}{3}$ **4** 3 kg, 33.6 N **5** 1.11 s, 4.67 m

6 437 N, 37.0 kg **7** $2.18\,\text{m}\,\text{s}^{-2}$

8 $1.09\,\text{m}\,\text{s}^{-2}$, 17.4 N **9** (a) $2.8\,\text{m}\,\text{s}^{-2}$
(b) 4.2 N (c) $1.67\,\text{m}\,\text{s}^{-1}$

10 0.813 s **11** (a) 0.715 m (b) $2.86\,\text{m}\,\text{s}^{-1}$
(c) 0.528 s **12** (a) $2.21\,\text{m}\,\text{s}^{-1}$ (b) 14.7 N
(c) particle, smooth surface, light inextensible string, smooth pulley

13 $1.09\,\text{m}\,\text{s}^{-2}$ 13.1N $0.656\,\text{ms}^{-2}$ down the plane

14 0.505 kg, 3.94 N **15** (a) $0.544\,\text{m}\,\text{s}^{-2}$
(b) 2.71 s (c) 2.16 m **16** 0.679, 1.05 m

Exercise 5E

1 (a) 0.2 N s (b) 20 N s (c) 13 500 N s
(d) 480 N s (e) 3×10^6 N s (f) 13 500 N s

2 12 000 N s **3** 16 200 N s **4** 2.6 N s

5 12 N s $24\,\text{m}\,\text{s}^{-1}$ **6** $4.5\,\text{m}\,\text{s}^{-1}$ **7** 18 N

8 (a) $6\mathbf{i}$ N s (b) $7.6\mathbf{i}\,\text{m}\,\text{s}^{-1}$

9 $-16\mathbf{j}$ N **10** 4 **11** $(8\mathbf{i} + 10\mathbf{j})$ N

12 5.4 N s **13** (a) 5.25 N s (b) 5.25 N s

14 $30\,\text{m}\,\text{s}^{-1}$ **15** (a) $7\,\text{m}\,\text{s}^{-1}$ (b) $5.94\,\text{m}\,\text{s}^{-1}$
(c) 5.18 N s

Exercise 5F

1 2 **2** 3 **3** 4

4 1.6 **5** 3 **6** 4

7 3 **8** 1 **9** 4

10 3 **11** $3\,\text{m}\,\text{s}^{-1}$ **12** $1.5\,\text{m}\,\text{s}^{-1}$

13 $6\,\text{m}\,\text{s}^{-1}$, 12 N s **14** $1.04\,\text{m}\,\text{s}^{-1}$

15 $4\,\text{m}\,\text{s}^{-1}$, **16** $500\,\text{m}\,\text{s}^{-1}$

17 $200\,\text{m}\,\text{s}^{-1}$ in the opposite direction

18 $6.67\,\text{m}\,\text{s}^{-1}$, 0.0533 m

Exercise 5G

1 (a) $3.92\,\text{m}\,\text{s}^{-2}$ (b) 1.5 m

2 (a) 0.206 (b) $1.01\,\text{m}\,\text{s}^{-2}$

3 (a) $2u$ (b) $16mu$ (c) final direction of motion of P.

4 (a) 2.64 N (b) 0.347 (c) 3.73 N (d) 2

5 (a) 0.8 (b) 0.48 N s

6 (a) 1.23 N (b) $2.45\,\text{m}\,\text{s}^{-2}$ (c) 7.35 m

7 (a) 51.2 N (b) 7.53 (c) 102.4 N (d) 6 m (e) tension in rope same on both sides of the pulley

8 (a) 2750 N (b) $\frac{1}{2}$ (c) 750 N (d) $1.8\,\text{m}\,\text{s}^{-2}$

9 (b) 66.3 m (c) 4.42 s (d) $10\,\text{m}\,\text{s}^{-1}$

Exercise 6A

1 (a) 12 N m anticlockwise
(b) 10 N m clockwise (c) 0 N m
(d) 15 N m clockwise (e) 10.6 N m clockwise
(f) 36.4 N m anticlockwise (g) 0 N m
(h) 61.3 N m anticlockwise

2 (a) 14 N m anticlockwise
(b) 6 N m clockwise
(c) 2 N m anticlockwise
(d) 7 N m anticlockwise (e) 0 N m
(f) 1.43 N m clockwise
(g) 15 Nm clockwise
(h) 5.07 Nm anticlockwise

3 10 N m anticlockwise

4 10 N m clockwise

5 4 N m anticlockwise

6 (a) 12 N m clockwise (b) 4 N m clockwise

7 (a) 16 N m anticlockwise
(b) 35 N m anticlockwise

8 (a) 4 N m anticlockwise
(b) 21 N m clockwise

Exercise 6B

1 (a) 6 N, 1.5 m (b) 6 N, 5 N (c) 2 N, 2 m (d) 3 N, 2 m (e) $2g$ N, $6g$ N (f) $6g$ N, 1 m

2 4 kg **3** $2\frac{1}{9}$ m **4** 32.7 N, 6.53 N

5 30.6 N, 28.2 N **6** 1.67 m

7 417 N, 368 N

8 (a) $100g$ N, $50g$ N
(b) $\frac{2}{3}$ m. Tree trunk is unlikely to be uniform. If centre of mass of the tree trunk were given, the model would be more accurate.

9 12 kg **10** $\frac{1}{3}$ m

11 $\frac{13}{4}d$; the weight acts at B

Review exercise 2

1 11° **2** 0.24 **3** $\frac{8}{27}$

4 (a) 0.577
(b) the book could be modelled as a particle.

5 (a) 11.4 N (b) 13.9 N

6 (a) 80 N (b) 69.3 N **7** 0.304

8 0.44

9 (a) −2, 7 (b) could be a particle.

10 (a) $(5\mathbf{i} - 12\mathbf{j})\,\text{m}\,\text{s}^{-2}$ $13\,\text{m}\,\text{s}^{-2}$ (b) 58.5 m

11 $0.52\,\text{m}\,\text{s}^{-2}$ **12** $3.4\,\text{m}\,\text{s}^{-2}$ 144°

13 0.58 **14** 1400 N

15 (a) $2.18\,\text{m}\,\text{s}^{-2}$ (b) 21.8 N (c) 4.36 m

16 (a) 3300 N (b) as particles
(c) no because all ropes are capable of being stretched when under tension.

17 (a) $4.2\,\text{m}\,\text{s}^{-2}$ (b) 3.36 N (c) $2.9\,\text{m}\,\text{s}^{-1}$
(d) 0.69 s

18 (a) $1.4\,\text{m}\,\text{s}^{-2}$ (b) 2.52 N (c) 3 s
(d) $4.2\,\text{m}\,\text{s}^{-1}$

19 (a) 8.4 N (b) 2.1 s (c) $0.35\,\text{m}\,\text{s}^{-2}$
(d) 9.45 N

20 (a) $1.375\,\text{m}\,\text{s}^{-2}$
(b) (i) T + 150 N ← Car → 2200 N
(ii) 200 N ← Caravan → T
(c) $1.16\,\text{m}\,\text{s}^{-2}$ (d) 952 N

21 (a) $1.4\,\text{m}\,\text{s}^{-2}$ (b) 3.36 N (c) $\frac{5}{7}$ s

22 0.371

23 (a) 0.42 N (b) 2.5

24 (a) 28 Ns
(b) could be modelled as a particles, gravitational force is constant, no air resistance or wind.

25 750 N

26 (a) 6.3 Ns

(b) can be modelled as a particle, gravitational force is constant, no air resistance or wind

27 560 kN

28 (b) 1580 Ns

(c) 6100 N

29 (b) 2.8 N

(c) 3.15 Ns

(d) 2.14 s

30 (a) 4 m s^{-1}

(b) 1

31 (a) 1

(b) 4 Ns

32 1.2 m s^{-1}

33 1.75 m

34 (a) 330 N, 706 N

(b) man as a particle, plank as a uniform rod

35 (a) 784 N

(b) 0.5 m

36 (a) 17.5 N

(b) 66°

(c) $\mathbf{P} = 3\mathbf{i} + 12\mathbf{j}$, $\mathbf{Q} = 4\mathbf{i} + 4\mathbf{j}$

37 17 N, 032°

39 2.2

40 0.75

41 (a) 43°

(b) 53 N

42 (b) $\sqrt{\dfrac{gh(1-3\mu)}{2}}$

(c) $\frac{1}{7} \leqslant \mu < \frac{1}{3}$

43 (b) 0.036 N s

44 (a) 0.5 m s^{-2}

(b) 0.58

(c) no

45 1.6d

Examination style paper M1

1 $q = 1, p = 2$

2 (a) $\dfrac{(2-k)}{k}u$ (b) $k < 2$ (c) $2Mu$

3 (a) 50 s (b) 20.6 m s^{-1}

4 (a) $r_A = (3\mathbf{i} + \mathbf{j}) + t(-\mathbf{i} + 5\mathbf{j})$

(b) $r_B = (-\mathbf{i} + 4\mathbf{j}) + t(2\mathbf{i} + 3\mathbf{j})$

(c) $(4 - 3t)\mathbf{i} + (-3 + 2t)\mathbf{j}$

(d) 3.20 p.m.

5 (a)

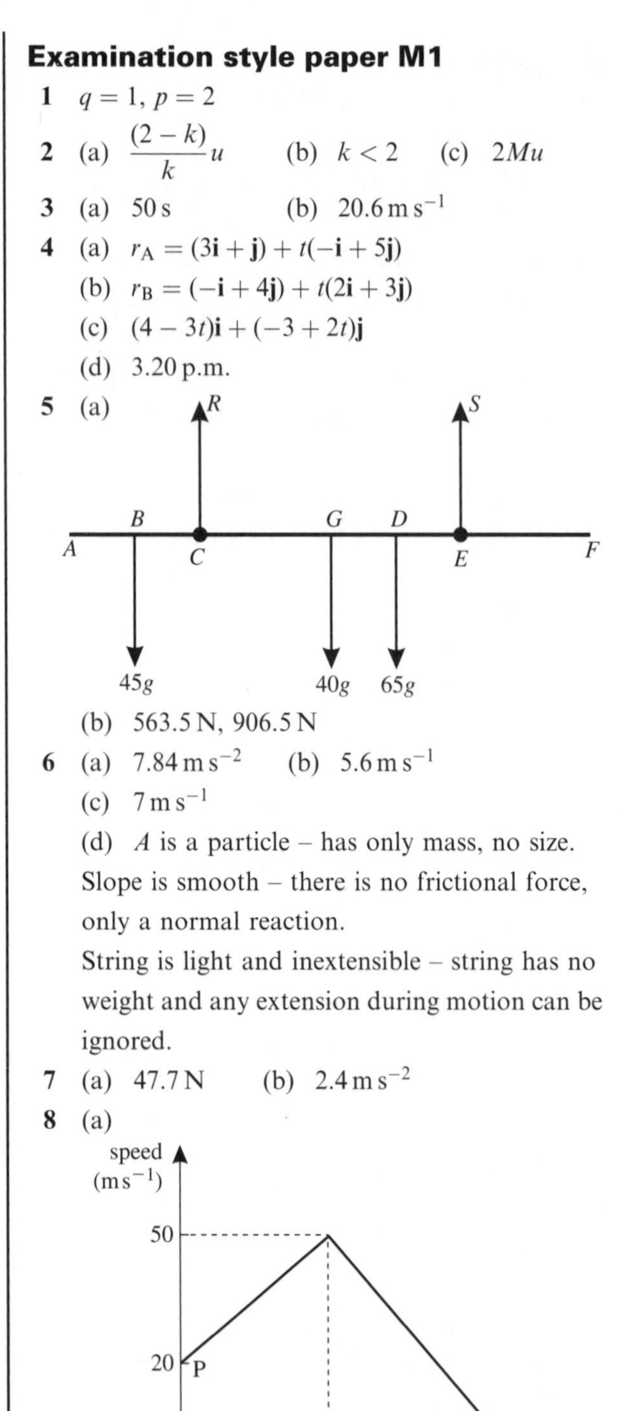

(b) 563.5 N, 906.5 N

6 (a) 7.84 m s^{-2} (b) 5.6 m s^{-1}

(c) 7 m s^{-1}

(d) A is a particle – has only mass, no size. Slope is smooth – there is no frictional force, only a normal reaction.
String is light and inextensible – string has no weight and any extension during motion can be ignored.

7 (a) 47.7 N (b) 2.4 m s^{-2}

8 (a)

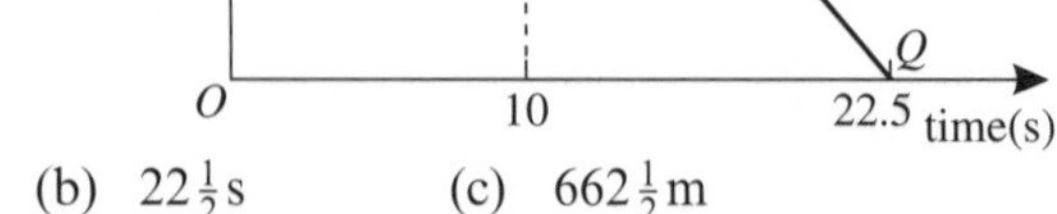

(b) $22\frac{1}{2}$ s (c) $662\frac{1}{2}$ m

List of symbols and notation

The following notation will be used in all Edexcel examinations.

$\in$	is an element of
$\notin$	is not an element of
$\{x_1, x_2, \ldots\}$	the set with elements $x_1, x_2, \ldots$
$\{x : \ldots\}$	the set of all x such that $\ldots$
$\mathrm{n}(A)$	the number of elements in set A
$\varnothing$	the empty set
$\mathscr{E}$	the universal set
A'	the complement of the set A
$\mathbb{N}$	the set of natural numbers, $\{1, 2, 3, \ldots\}$
$\mathbb{Z}$	the set of integers, $\{0, \pm 1, \pm 2, \pm 3, \ldots\}$
$\mathbb{Z}^+$	the set of positive integers, $\{1, 2, 3, \ldots\}$
$\mathbb{Z}_n$	the set of integers modulo n, $\{0, 1, 2, \ldots, n-1\}$
$\mathbb{Q}$	the set of rational numbers $\left\{\frac{p}{q} : p \in \mathbb{Z}, q \in \mathbb{Z}^+\right\}$
$\mathbb{Q}^+$	the set of positive rational numbers, $\{x \in \mathbb{Q} : x > 0\}$
$\mathbb{Q}_0^+$	the set of positive rational numbers and zero, $\{x \in \mathbb{Q} : x \geqslant 0\}$
$\mathbb{R}$	the set of real numbers
$\mathbb{R}^+$	the set of positive real numbers, $\{x \in \mathbb{R} : x > 0\}$
$\mathbb{R}_0^+$	the set of positive real numbers and zero, $\{x \in \mathbb{R} : x \geqslant 0\}$
$\mathbb{C}$	the set of complex numbers
(x, y)	the ordered pair x, y
$A \times B$	the cartesian product of sets A and B, $A \times B = \{(a, b) : a \in A, b \in B\}$
$\subseteq$	is a subset of
$\subset$	is a proper subset of
$\cup$	union
$\cap$	intersection
$[a, b]$	the closed interval, $\{x \in \mathbb{R} : a \leqslant x \leqslant b\}$
$[a, b)$, $[a, b[$	the interval $\{x \in \mathbb{R} : a \leqslant x < b\}$
$(a, b]$, $]a, b]$	the interval $\{x \in \mathbb{R} : a < x \leqslant b\}$
(a, b), $]a, b[$	the open interval $\{x \in \mathbb{R} : a < x < b\}$
$y R x$	y is related to x by the relation R
$y \sim x$	y is equivalent to x, in the context of some equivalence relation
$=$	is equal to
$\neq$	is not equal to
$\equiv$	is identical to *or* is congruent to

$\approx$	is approximately equal to
$\cong$	is isomorphic to
$\propto$	is proportional to
$<$	is less than
$\leqslant, \ngtr$	is less than or equal to, is not greater than
$>$	is greater than
$\geqslant, \nless$	is greater than or equal to, is not less than
∞	infinity
$p \wedge q$	p and q
$p \vee q$	p or q (or both)
$\sim p$	not p
$p \Rightarrow q$	p implies q (if p then q)
$p \Leftarrow q$	p is implied by q (if q then p)
$p \Leftrightarrow q$	p implies and is implied by q (p is equivalent to q)
$\exists$	there exists
$\forall$	for all
$a + b$	a plus b
$a - b$	a minus b
$a \times b$, ab, $a.b$	a multiplied by b
$a \div b$, $\frac{a}{b}$, a/b	a divided by b
$\sum_{i=1}^{n} a_i$	$a_1 + a_2 + \ldots + a_n$
$\prod_{i=1}^{n} a_i$	$a_1 \times a_2 \times \ldots \times a_n$
$\sqrt{a}$	the positive square root of a
$\lvert a \rvert$	the modulus of a
$n!$	n factorial
$\binom{n}{r}$	the binomial coefficient $\frac{n!}{r!(n-r)!}$ for $n \in \mathbb{Z}^+$; $\frac{n(n-1)\ldots(n-r+1)}{r!}$ for $n \in \mathbb{Q}$
$\mathrm{f}(x)$	the value of the function f at x
$\mathrm{f}: A \to B$	f is a function under which each element of set A has an image in set B
$\mathrm{f}: x \mapsto y$	the function f maps the element x to the element y
f^{-1}	the inverse function of the function f
$\mathrm{g} \circ \mathrm{f}$, gf	the composite function of f and g which is defined by $(\mathrm{g} \circ \mathrm{f})(x)$ or $\mathrm{gf}(x) = \mathrm{g}(\mathrm{f}(x))$
$\lim_{x \to a} \mathrm{f}(x)$	the limit of $\mathrm{f}(x)$ as x tends to a
Δx, δx	an increment of x
$\frac{\mathrm{d}y}{\mathrm{d}x}$	the derivative of y with respect to x
$\frac{\mathrm{d}^n y}{\mathrm{d}x^n}$	the nth derivative of y with respect to x

$f'(x), f''(x), \ldots f^{(n)}(x)$	the first, second, ... nth derivatives of $f(x)$ with respect to x
$\int y \, dx$	the indefinite integral of y with respect to x
$\int_a^b y \, dx$	the definite integral of y with respect to x between the limits $x = a$ and $x = b$
$\frac{\partial V}{\partial x}$	the partial derivative of V with respect to x
$\dot{x}, \ddot{x}, \ldots$	the first, second, . . . derivatives of x with respect to t
e	base of natural logarithms
e^x, exp x	exponential function of x
$\log_a x$	logarithm to the base a of x
$\ln x$, $\log_e x$	natural logarithm of x
$\lg x$, $\log_{10} x$	logarithm of x to base 10
sin , cos , tan cosec, sec, cot	the circular functions
arcsin, arccos, arctan arccosec, arcsec, arccot	the inverse circular functions
sinh, cosh, tanh cosech, sech, coth	the hyperbolic functions
arsinh, arcosh, artanh, arcosech, arsech, arcoth	the inverse hyperbolic functions
i, j	square root of -1
z	a complex number, $z = x + iy$
Re z	the real part of z, Re $z = x$
Im z	the imaginary part of z, Im $z = y$
$\|z\|$	the modulus of z, $\|z\| = \sqrt{(x^2 + y^2)}$
arg z	the argument of z, arg $z = \arctan \frac{y}{x}$
z^*	the complex conjugate of z, $x - iy$
M	a matrix **M**
$\mathbf{M}^{-1}$	the inverse of the matrix **M**
$\mathbf{M}^{T}$	the transpose of the matrix **M**
det **M**, $\|\mathbf{M}\|$	the determinant of the square matrix **M**
a	the vector **a**
$\overrightarrow{AB}$	the vector represented in magnitude and direction by the directed line segment AB
$\hat{\mathbf{a}}$	a unit vector in the direction of **a**
i, j, k	unit vectors in the directions of the cartesian coordinate axes
$\|\mathbf{a}\|$, a	the magnitude of **a**
$\|\overrightarrow{AB}\|$, AB	the magnitude of $\overrightarrow{AB}$
a . b	the scalar product of **a** and **b**
$\mathbf{a} \times \mathbf{b}$	the vector product of **a** and **b**

A, B, C, etc	events
$A \cup B$	union of the events A and B
$A \cap B$	intersection of the events A and B
$\mathrm{P}(A)$	probability of the event A
A'	complement of the event A
$\mathrm{P}(A\|B)$	probability of the event A conditional on the event B
X, Y, R, etc.	random variables
x, y, r, etc.	values of the random variables X, Y, R, etc
$x_1, x_2 \ldots$	observations
$f_1, f_2, \ldots$	frequencies with which the observations $x_1, x_2, \ldots$ occur
$\mathrm{p}(x)$	probability function $\mathrm{P}(X = x)$ of the discrete random variable X
$p_1, p_2, \ldots$	probabilities of the values $x_1, x_2, \ldots$ of the discrete random variable X
$\mathrm{f}(x), \mathrm{g}(x), \ldots$	the value of the probability density function of a continuous random variable X
$\mathrm{F}(x), \mathrm{G}(x), \ldots$	the value of the (cumulative) distribution function $\mathrm{P}(X \leqslant x)$ of a continuous random variable X
$\mathrm{E}(X)$	expectation of the random variable X
$\mathrm{E}[\mathrm{g}(X)]$	expectation of $\mathrm{g}(X)$
$\mathrm{Var}(X)$	variance of the random variable X
$\mathrm{G}(t)$	probability generating function for a random variable which takes the values 0, 1, 2, ...
$\mathrm{B}(n, p)$	binomial distribution with parameters n and p
$\mathrm{N}(\mu, \sigma^2)$	normal distribution with mean μ and variance σ^2
μ	population mean
σ^2	population variance
σ	population standard deviation
$\bar{x}$, m	sample mean
s^2, $\hat{\sigma}^2$	unbiased estimate of population variance from a sample, $s^2 = \frac{1}{n-1}\sum(x_i - \bar{x})^2$
ϕ	probability density function of the standardised normal variable with distribution N(0, 1)
Φ	corresponding cumulative distribution function
ρ	product-moment correlation coefficient for a population
r	product-moment correlation coefficient for a sample
$\mathrm{Cov}(X, Y)$	covariance of X and Y

Index